Totality

Totality
Eclipses of the Sun

Mark Littmann and Ken Willcox

Foreword by Donald N. B. Hall
Director, Institute for Astronomy
University of Hawaii

University of Hawaii Press / Honolulu

Printed in the United States of America
91 92 93 94 95 96 5 4 3 2 1

Library of Congress Cataloging-in-Publication Data

Littmann, Mark, 1939–
Totality : eclipses of the sun /
Mark Littmann and Ken Willcox.
p. cm.
Includes bibliographical references and index.
ISBN 0-8248-1371-5 (acid-free paper)
1. Eclipses, Solar. I. Willcox, Ken, 1943–
II. Title.
QB541.L69 1991
523.7'8—dc20 90-23823
CIP

Cover illustration: Corona and prominences.
February 26, 1979. Photo © 1979 by Jay Anderson

University of Hawaii Press books are printed
on acid-free paper and meet the guidelines
for permanence and durability of the Council
on Library Resources

From Mark Littmann

To my mother, Muriel Stein Littmann,
and in memory of my father, Lewis Littmann, M.D.

From Ken Willcox

To my wife, Sara

Contents

Foreword

Total eclipses of the Sun draw solar astronomy enthusiasts to the far corners of the globe. Amateurs are attracted by expectations of the breathtaking beauty of the eclipsed Sun during the few minutes of totality. Professional astronomers are driven to get optimum amounts of research done during those same few minutes, while trying not to be too distracted by the eclipse's stark spectacle.

The path of complete darkness swept out by the Moon's shadow during a total eclipse is typically only about 100 miles across. Moving across the surface of the Earth at more than twice the speed of sound, the shadow passes any point along the path in minutes. The eclipse track can be predicted with great accuracy and observers agonize over which points along it offer the best viewing positions with the least threat of clouds. This usually results in observers selecting remote mountain sites, desert areas, or isolated islands from which to view the eclipse. Solar astronomers often invest months, sometimes years, in preparing for the event. They spend large sums of money transporting complex equipment to these remote sites and doing their utmost to ensure it will work perfectly on the day of the eclipse. The possibility of being clouded out, or worse—making some simple error that prevents good data being taken—strikes terror into most who have undertaken such endeavors. Mistakes as simple as failing to remove lens caps or filters happen surprisingly often under the stress of the eclipse.

Other pitfalls are far more difficult to predict. The great inventor Thomas Alva Edison traveled to the plains of Wyoming for the eclipse of July 1878, planning to make observations of the eclipsed Sun in the infrared using his latest discovery, the tassimeter. The persistent prairie winds relentlessly shook his unprotected small telescope until, in desperation, he moved to a hen house where he was able to prepare for totality. However, his plans were thwarted when, as the darkness of eclipse deepened, the resident hens returned to roost in great confusion around his experiment.

Expeditions to eclipses often become great adventures. On May 30, 1965, a major eclipse was to sweep across the Pacific from New

Zealand to Peru. U.S. astronomers chartered the sleek, 160-foot schooner *Goodwill* to sail from Los Angeles to the tiny, tropical islands of Bellingshausen and Manuae, near Tahiti. The *Goodwill* stopped in Papeete to take on board some 40 scientists, including John Jefferies and Frank Orrall from the fledgling solar astronomy program at the University of Hawaii. The eclipse was observed successfully, but I have heard many reminiscences about rot in the *Goodwill*'s hull, unruly crew, an erratic compass, and fierce storms. Nostalgia abounds for the days of tranquility experienced on an isolated South Seas atoll!

My own eclipse career started out from great heights. I was one of a group of seven scientists privileged to observe the solar eclipse of 1973 from the French prototype supersonic aircraft, Concorde, high above the Sahara Desert. Able to fly at more than twice the speed of sound, the Concorde was just able to keep pace with the Moon's shadow and so remained within totality for more than an hour. The French company Aerospatiale generously provided the aircraft and the engineering support to retrofit windows into the roof of Concorde 001 to allow scientific observations during the eclipse. After a busy spring in the south of France and a hectic deployment to La Palma—the jumping off point in the Canary Islands—the aircraft and crew performed flawlessly on the day of the eclipse. I still remember clearly the striking, visual spectacle as we intercepted the Moon's shadow over the western Sahara and rode with it for nearly an hour-and-a-quarter. At the 70,000-foot altitude at which the Concorde flew, one could see out into the penumbra, producing the eerie effect of a desert thunderstorm.

The eclipse of 1991 represents a once-in-a-lifetime opportunity for professional astronomers. Instead of the astronomers going to the eclipse, the eclipse has come to the mountain—in this case, Mauna Kea, which boasts the largest collection of astronomical light-gathering power of any observatory on the globe. The center line of totality passes within a mile of the observatories when atmospheric distortion of images is at a minimum. Instead of astronomers having to "make do" with small telescopes and equipment which can be transported to remote sites, the eclipse of 1991 will allow us to aim the world's most advanced telescopes and their sophisticated auxiliary equipment on the eclipsed Sun. The enormous light-gathering power of the telescopes on Mauna Kea and the detailed images possible from the site will allow a new class of experiments to probe the detailed structure of the Sun's chromosphere and extended atmosphere, the corona, and to study the acceleration of the solar wind. Mauna Kea's superb infrared qualities and the availability of the sophisticated infrared cameras at the site have led many scientists to plan detailed studies of heated dust in the vicinity of the Sun. The total eclipse of 1991 promises a bonanza of new scientific results.

Mark Littmann and Ken Willcox have produced a valuable and entertaining work, which provides the general reader with a historical account of eclipses from the earliest times through recent scientific research together with a clear description of the mechanics of eclipses and their place in modern solar research.

DONALD N. B. HALL
Director
Institute for Astronomy
University of Hawaii

Acknowledgments

This book—perhaps no book—is the work of the authors alone. There are very special people who, by their wealth of knowledge, their creativity, and their graciousness made the completion of our book possible. To these people, there is no truly adequate way to express appreciation. We can only try to single out some of them so that you too can know who helped us so greatly.

Ruth S. Freitag of the Library of Congress was absolutely indispensible. Extraordinarily knowledgeable in the history of astronomy, she supplied references, identified little-known scientists of the past, found pictures, submitted to an interview on her eclipse experiences, and reviewed the entire manuscript.

A remarkable range of vital contributions came from Charles A. Lindsey, a distinguished solar physicist at the Institute for Astronomy, University of Hawaii. Not only did he review the manuscript, he also created most of the book's diagrams (rendered camera ready by Wendy Nakano) and patiently endured interviews on the use of eclipses in his solar research.

Dennis di Cicco of *Sky & Telescope* also reviewed a draft of the book and contributed some of his superb photography. We profited enormously from his, Charlie's, and Ruth's expertise and wise advice.

In our research, we consulted experts in a wide variety of fields. Some generously contributed vignettes to our book, giving it a richness it would not otherwise have. We are deeply grateful to Lucian V. Del Priore, M.D., Stephen J. Edberg, Alan D. Fiala, Carl Littmann, Laurence A. Marschall, and Jay M. Pasachoff.

A special note of thanks to John R. Beattie. As he recounted his observing experiences, he provided such a ringing moment-by-moment description of a total eclipse that it became the nucleus of our first chapter.

For this book, we interviewed some of the most experienced eclipse veterans from the ranks of both professional and amateur astronomers. Their valuable perspectives form the heart of the chapters on observing a total eclipse, modern scientific uses for solar eclipses, and eclipse photography. Often they kindly contributed their eclipse photographs for use in this book and reviewed these chapters to reduce our errors. Their names are mentioned in the chapters, but they deserve additional recognition and gratitude here: Jay Anderson, Eric Becklin, Richard Berry, David W. Dunham, Richard Fisher, William C. Livingston, George Lovi, Frank Orrall, Leif J. Robinson, Virginia Roth, Gary Rottman, and Jack B. Zirker. Each of these experts suggested others who have contributed significantly to public appreciation of eclipses or to solar research. We regret that we did not have time to interview all the veterans they suggested or that we intended.

Crucial to a book on solar eclipses are maps of eclipse paths. We had the help of two exceptional eclipse computers and cartographers: Fred Espenak and Jean Meeus.

There are other eclipse veterans and specialists who kindly helped us with information and ideas: Anthony F. Aveni, Kenneth Brecher, Andrew P. Fraknoi, Kevin Krisciunas, Robert S. Harrington, Sabatino Sofia, and E. Myles Standish, Jr.

Two science writers graciously reviewed the whole manuscript to help us improve our presentation. Special thanks to Beatrice C. Owens and Ann L. Rappoport.

Without great library services, research for books such as this is impossible. Enormous gratitude to the Loyola/Notre Dame Library in Baltimore. A more talented and gracious staff is hard to imagine. We wish we could list all the librarians who helped so significantly with this project. We are also grateful to the U.S. Naval Observatory Library for allowing so many of its books to migrate to us for this project. Thanks too to Gerry Grimm of Baltimore and to John Carper of the John G. Wolbach Library, Harvard College Observatory, for their contributions.

Eclipse lore crosses all boundaries, and we needed help from careful and fast-working translators. Thanks to David and Esther Littmann of Detroit and to Geoffrey K. Gay and Richard E. McCarron of Loyola College.

Of course, a book does not appear without the faith, encouragement, and insight of a publisher. We are deeply grateful to the skillful team at the University of Hawaii Press.

M.L. and K.W.

I am profoundly grateful to Peggy, my wife, without whose love, self-sacrifice, insight, and encouragement this project would have been impossible. To Beth and Owen, thanks for their interest and understanding and love throughout this project. I would also like to thank Jane Littmann, Tom Owens, and Paul Rappoport for their help through the writing of this book.

To Ken Willcox, my admiration for his courage, dedication, and talent. It is a pleasure and honor to work with him.

This book was written while I was teaching astronomy at Loyola College in Baltimore. I am very grateful to Helene Perry, former chairman of the Physics Department, and to all the members of the Physics faculty and staff for their encouragement and friendship.

In the midst of this project, I received a remarkable offer from the School of Journalism at the University of Tennessee. To Professors George Everett, James Crook, and my new colleagues in the College of Communications, I am deeply honored that you have invited me to join you.

M.L.

Without the support, encouragement, and understanding love of my wife, Sara, my participation in this project would not have been possible. She is a very big part of what I am. I am profoundly grateful to my father, Delbert Willcox, who, when I was young, taught me much of what I know about photography; to my mother, Madelyn Burgess Willcox, who encouraged my interest in science and astronomy; and to my late aunt, Helen Burgess, who gave me a 3-inch Moonscope for Christmas in 1955 that stimulated my curiosity about God's universe.

Special thanks to Phillips Petroleum Company, my employer, for dedication to improving science education in America, and to the Astronomical League for the wealth of opportunities I have received as a member and for the support I have received during my terms as vice president and president.

This book was written during a most difficult time in my life: throughout its writing I was undergoing treatment for cancer. The attention I was able to give the book is a direct result of unselfish love provided by Dr. and Mrs. Malcolm Granberry of Houston, who hold a special place in the lives of my family. The most credit for this book, and for my participation in it, belongs to Mark Littmann, whose gracious and professional guidance provided a challenge that helped lift me above my fears of what the future might hold.

Through it all has come a fuller realization that people are the most important creation in the universe, and that my Lord has put me in contact with some of His best.

K.W.

Some people see a partial eclipse and wonder why others talk so much about a total eclipse. Seeing a partial eclipse and saying that you have seen an eclipse is like standing outside an opera house and saying that you have seen the opera; in both cases, you have missed the main event.

Jay M. Pasachoff (1983)

1

The Experience of Totality

First contact. A tiny nick appears on the western side of the Sun.[1] The eye detects no difference in the amount of sunlight. Nothing but that nick portends anything out of the ordinary. But as the nick becomes a gouge in the face of the Sun, a sense of anticipation begins. This will be no ordinary day.

Still, the pace is leisurely for the first half hour or so, until the Sun is more than half covered. Now, gradually at first, then faster and faster, extraordinary things begin to happen. The sky is still bright, but the blue is a little duller. On the ground around you the light is beginning to diminish. Over the next 10 to 15 minutes, the landscape takes on a steely gray metallic cast.

As the minutes pass, the pace quickens. With about a quarter hour left until totality, the western sky is darker than the eastern, regardless of where the Sun is in the sky. The shadow of the Moon is approaching. Even if you have never seen a total eclipse of the Sun before, you know that something amazing is going to happen, is happening now—and that it is beyond normal human experience.

Less than 15 minutes until totality. The Sun, a narrowing crescent, is still fiercely bright, but the blueness of the sky has deepened into blue-gray or violet. The darkness of the sky begins to close in around the Sun. The Sun does not fill the heavens with brightness anymore.

Early partial phase of a total eclipse (June 11, 1983). © 1983 Jay M. Pasachoff

Five minutes to totality. The darkness in the west is very noticeable and gathering strength, a dark amorphous form rising upward and spreading out along the western horizon. It builds like a massive storm, but in utter silence, with no rumble of distant thunder. Then the darkness begins to float up above the horizon, revealing a yellow or orange twilight beneath. You are already seeing through the Moon's narrow shadow to the resurgent sunlight beyond.

The acceleration of events intensifies. The crescent Sun is now a blazing white sliver, like a welder's torch. The darkening sky continues to close in around the Sun, faster, engulfing it. Minutes have become seconds. The ends of the bare sliver of the Sun break into individual dots of intense white light—Baily's Beads—the last rays of sunlight passing through the deepest lunar valleys. Opposite the beads, a ghostly round silhouette looms into view. It is the dark limb of the Moon, framed by a white opalescent glow that creates a halo around the darkened Sun. The corona, the most striking and unexpected of all the features of a total eclipse, is emerging.

Along the shrinking sliver of the Sun, the beads flicker, each

Diamond Ring Effect (February 16, 1980). *NOAO*

lasting but an instant and vanishing as new ones form. And now there is only one, set like a single diamond in a ring. The remaining one small dot of sunlight fades as if it were sucked into an abyss. Totality.

Where the Sun once stood, there is a black disk in the sky, outlined by the soft pearly white glow of the corona, about the brightness of a Full Moon. Small but vibrant reddish features stand at the eastern rim of the Moon's disk, contrasting vividly with the white of the corona and the black where the Sun is hidden. These are the prominences, giant clouds of hot gas in the Sun's lower atmosphere. They are always a surprise, each unique in shape and size, different yesterday and tomorrow from what they are at this special moment.

You are standing in the shadow of the Moon. It is dark enough to see Venus and Mercury and whichever of the brightest planets and stars happen to be close to the Sun's position and above the horizon. But it is not the dark of night. Looking across the landscape at the horizon in all directions, you see beyond the shadow to where the eclipse is not total, an eerie twilight of orange and yellow. From this

Corona (March 7, 1970). *NASA*

light beyond the darkness that envelops you comes an inexorable sense that time is limited.

Now, at the midpoint in totality, the corona stands out most clearly, its shape and extent never quite the same from one eclipse to another. Only the eye can do the corona justice, its special pattern of faint wisps and spikes on this day never seen before and never to be seen again. Yet around you at the horizon is a warning that totality is drawing to an end. The west is brightening while in the east the darkness is deepening and descending toward the horizon. Above you, prominences appear at the western edge of the Moon. The edge brightens.

Suddenly totality is over. A dot of sunlight appears. Quickly, this heavenly diamond broadens into a band of several jewels and then a sliver of the crescent Sun once more. The dark shadow of the Moon silently slips past you and rushes off toward the east. It is then you ask, "When is the next one?"[2]

> If God had consulted me before embarking upon creation, I would have recommended something simpler.
>
> Alfonso X, King of Castile (1252)

2

The Great Celestial Cover-up

A total eclipse of the Sun is exciting and even profoundly moving.

But what causes a total solar eclipse? The Moon blocks the Sun from view. And that is all you absolutely need to know to enjoy a solar eclipse. So you can now skip to the next chapter.

If however you are reading this paragraph, you are right: there is more to tell—about dark shadows and oblong orbits and tilts and danger zones and amazing coincidences. Yet before you venture further, promise yourself that if your eyes begin to glaze over, you will stop reading this chapter immediately and go on to the next. You must not let celestial mechanics, or our explanation of it, stand in the way of your enjoyment of the wild, wacky, and wonderful things people have thought and done about solar eclipses.

Moon Plucking

How big is the Moon in the sky? What is its angular size? Use your thumb and index finger to make the shape of a "C." (For right handers, the "C" will be facing backwards.) Keep that shape and extend your arm upward as far from your body as possible. Imagine that you are trying to pluck the Moon out of the sky ever so carefully,

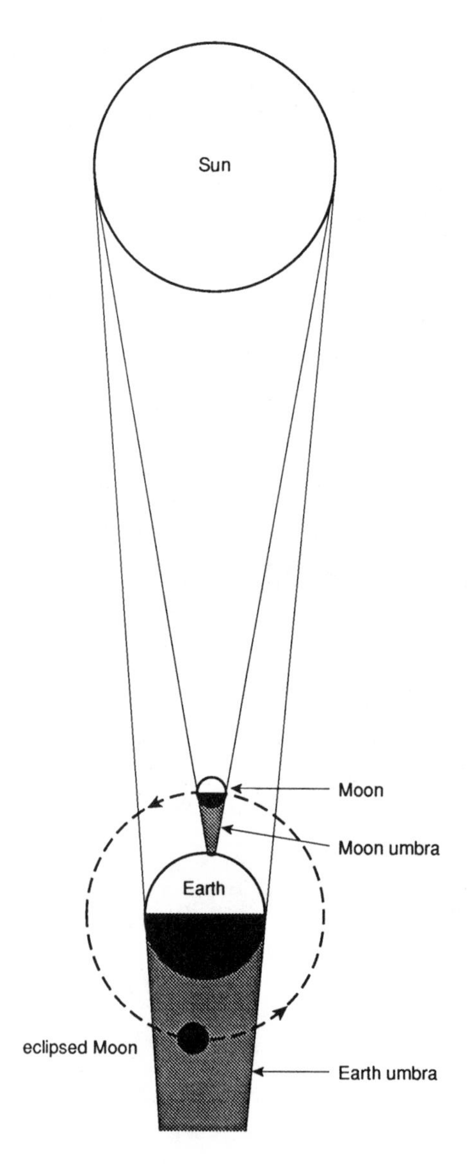

A lunar eclipse occurs when the Moon passes into the Earth's shadow (umbra). A total solar eclipse occurs when the Moon's umbra touches the Earth. The relative sizes and distances of the Sun, Moon, and Earth in this diagram are not to scale.

squeezing down with your thumb and index finger until you are just barely touching the top and bottom of the Moon, trapping the Moon between your fingers. How big is it? The size of a grape? A peach? An orange? It is the size of a pea. (You can win bets at cocktail parties with this question.) The Moon has an angular size of 0.5 degree.

Total eclipse at Fryeburg, Maine (August 31, 1932). *Lick Observatory*

How large is the Sun in the sky? Your friends will almost all immediately guess that it is bigger. Before they damage their eyes by trying the Moon pinch on the Sun, just remind them that a total eclipse is caused by the Moon completely covering the Sun, so the Sun must be no bigger than a pea in the sky as well. It is the brightness of the Moon and especially the Sun that deceives people into overestimating their angular sizes.

Now that you have collected on your bets and can lead a life of leisure, think about the remarkable coincidence that allows us to have total eclipses of the Sun. The Sun is 400 times the diameter of the Moon, yet it is about 400 times farther from the Earth, so the two appear almost the same size in the sky, thereby providing the possibility of total eclipses. If the Moon, 2,160 miles (3,476 km) in diameter, were 161 miles (260 km) smaller than it is, or if it were farther away so that it appeared smaller, people on Earth would never see a total eclipse.[1]

It is amazing that there are total eclipses of the Sun at all. As it is, total eclipses can just barely happen. The Sun is not always exactly the same angular size in the sky. The reason is that the Earth's orbit is

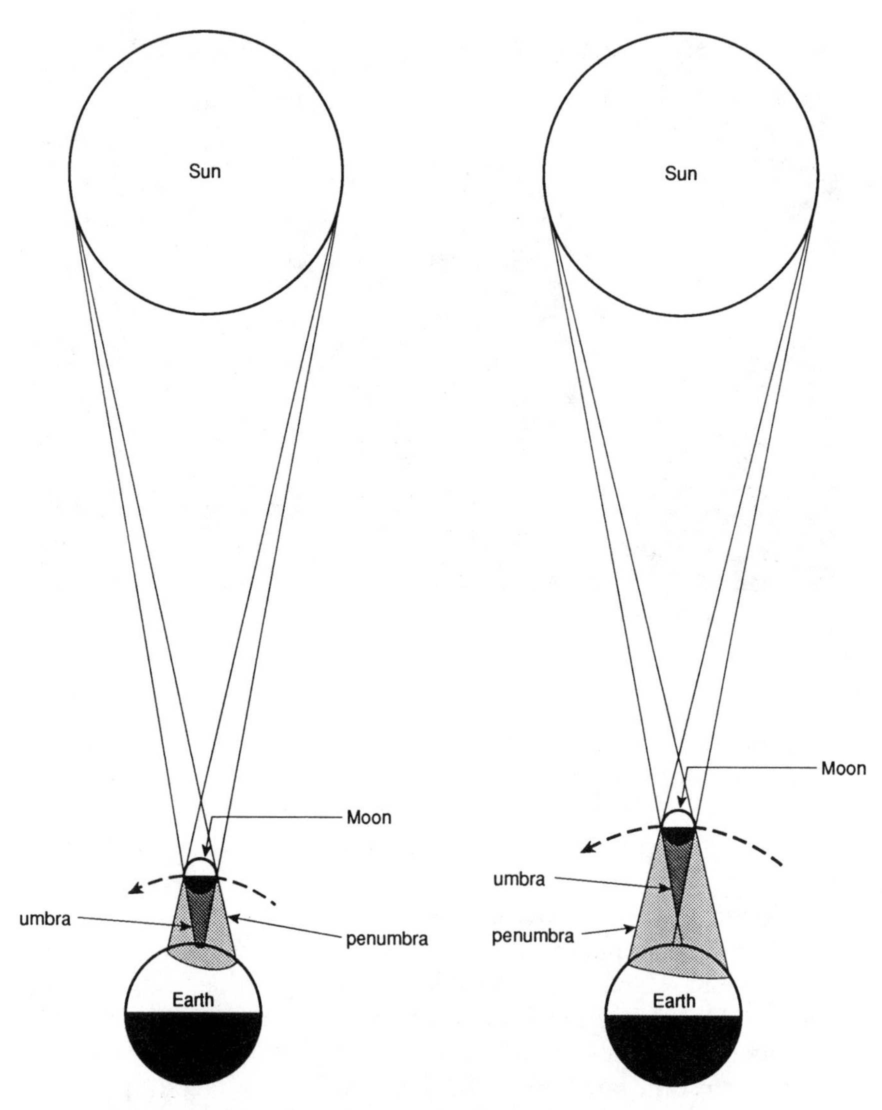

Configuration of a total solar eclipse *(left)* and an annular eclipse *(right)*. Within the Moon's umbra (dark converging cone), the entire surface of the Sun is blocked from view. In the penumbra (lighter diverging cone), a fraction of the sunlight is blocked, resulting in a partial eclipse. When the Moon's umbra ends in space *(right)*, a total eclipse does not occur. Projecting the cone through the tip of the umbra onto the Earth's surface defines the region in which an annular eclipse is seen.

not circular but elliptical, so the Earth's distance from the Sun varies. When the Earth is closest to the Sun (in early January),[2] the Sun's disk is slightly larger in angular diameter, and it is harder for the Moon to cover the Sun to create a total eclipse.

An even more powerful factor is the Moon's elliptical orbit around the Earth. When the Moon is its average distance from the Earth or farther, its disk is too small to occult the Sun completely. In the midst of such an eclipse, a circle of brilliant sunlight surrounds the Moon, giving the event a ringlike appearance; hence the name *annular* eclipse (from the Latin *annulus*, meaning "ring"). Because the angular diameter of the Moon is smaller than the angular diameter of the Sun on the average, annular eclipses are more frequent than total eclipses.

But the Moon does not just dangle motionless in front of the Sun. It is in orbit around the Earth. It catches up with and passes the Sun's position in the sky about once a month (a period of time derived from this circuit of the Moon). The actual time for the Moon to complete this cycle is 29.53 days, and it is called a *synodic month,*

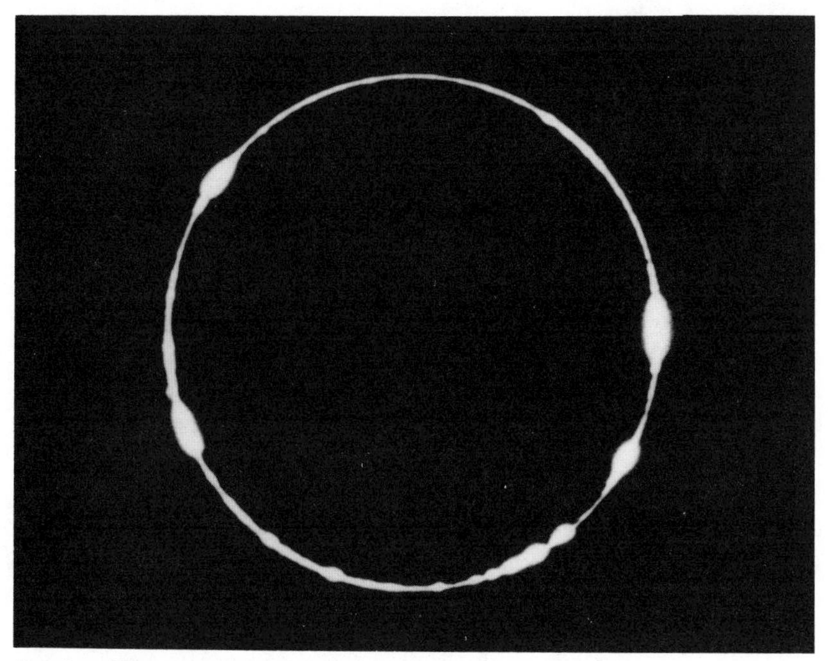

Annular eclipse, with Baily's Beads. *Lick Observatory*

The Shadow of the Moon

	maximum miles (km)	minimum miles (km)	mean miles (km)
Moon's distance from Earth (center to center)	252,710 (406,700)	221,460 (356,400)	238,900 (384,500)
Length of Moon's shadow cone (umbra)	236,700 (381,000)	226,800 (365,000)	231,100 (372,000)

Significance

On the *average*, the Moon's shadow is not long enough to reach the Earth, therefore total eclipses are less frequent than annular eclipses.

Angular Size of the Sun and Moon

	maximum	minimum	mean
Angular diameter of the Moon as seen from Earth	33′30″	29′22″	31′05.16″
Angular diameter of the Sun as seen from Earth	32′36″	31′32″	31′59.26″

Significance

The angular diameter of the Moon can *exceed* the angular diameter of the Sun by as much as 6.2 percent or almost 2 arc minutes, producing a total eclipse.

The angular diameter of the Sun can *exceed* the angular diameter of the Moon by as much as 11 percent or almost 3¼ arc minutes, producing an annular eclipse.

On the *average*, the angular diameter of the Moon is smaller than the angular diameter of the Sun, so total eclipses are less frequent than annular eclipses.

after the Greek *synodos*, "meeting"—the meeting of the Sun and the Moon. Because the Moon gives off no light of its own and shines only by reflected sunlight, its orbit around the Earth changes its angle to the Sun and determines its phase. In 29.53 days, the Moon goes from New Moon through Full Moon and back to New Moon again. This period is called a lunar month, *lunation*. Solar eclipses can take place only at New Moon (dark-of-the-Moon) and lunar eclipses may occur only at Full Moon.

So why do we not have an eclipse of the Sun every 29.53 days—every time the Moon passes the Sun's position? The reason is that the Moon's orbit around the Earth is tilted to the Earth's orbit around

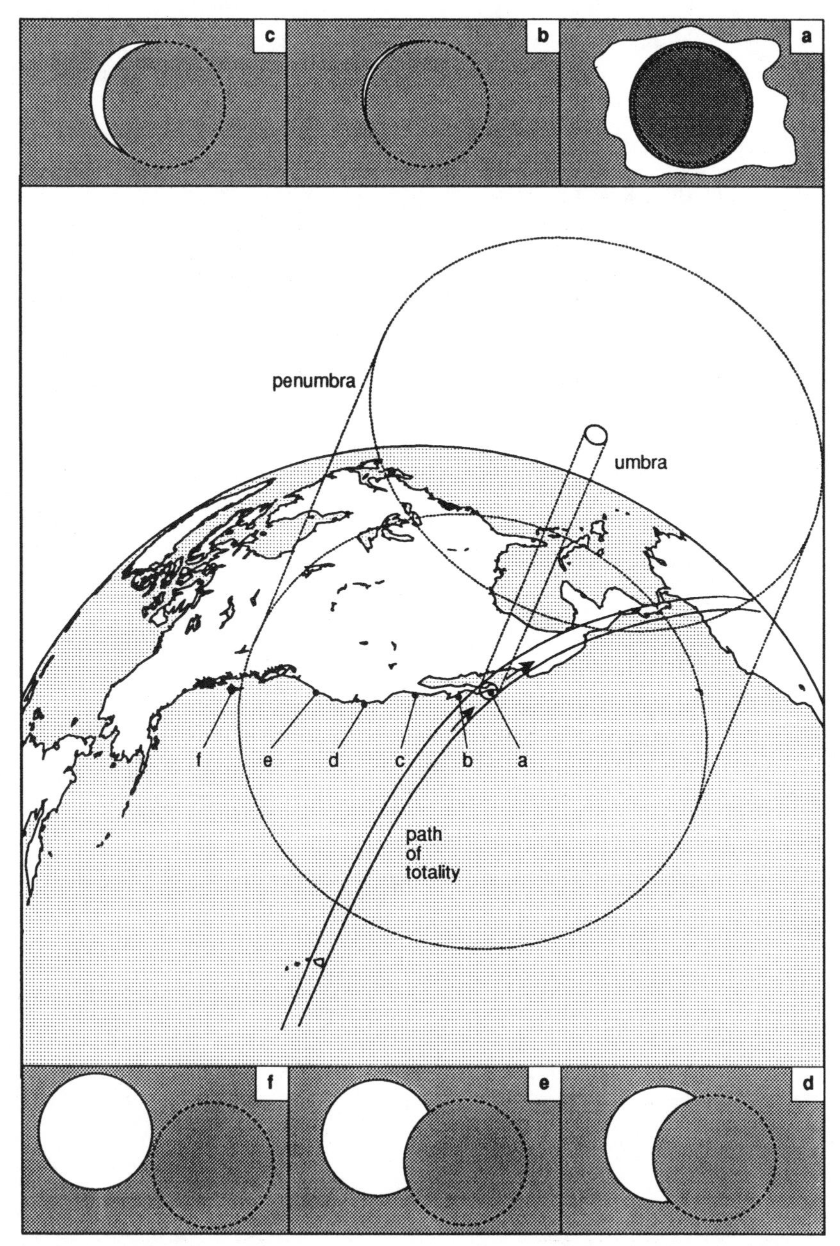

During a total eclipse of the Sun, the tip of the Moon's shadow touches the Earth and the Moon's orbital velocity carries the shadow rapidly eastward. Only along a narrow path is the eclipse total. Regions to the side of the path of totality experience varying degrees of partial eclipse. (This diagram is accurate for 11:50 A.M. PDT during the total solar eclipse of July 11, 1991.)

the Sun by about 5 degrees, so that the Moon usually passes above or below the Sun's position in the sky and cannot block the Sun from our view.

"Danger Zones"

The Moon's tilted orbit crosses the Earth's orbit at two places. Those intersections are called *nodes*. Node is from the Latin word meaning "knot," in the sense of weaving, where two threads are tied together. The point at which the Moon crosses the Earth's orbit going northward is the ascending node. Going south, the Moon crosses the Earth's orbit at the descending node.

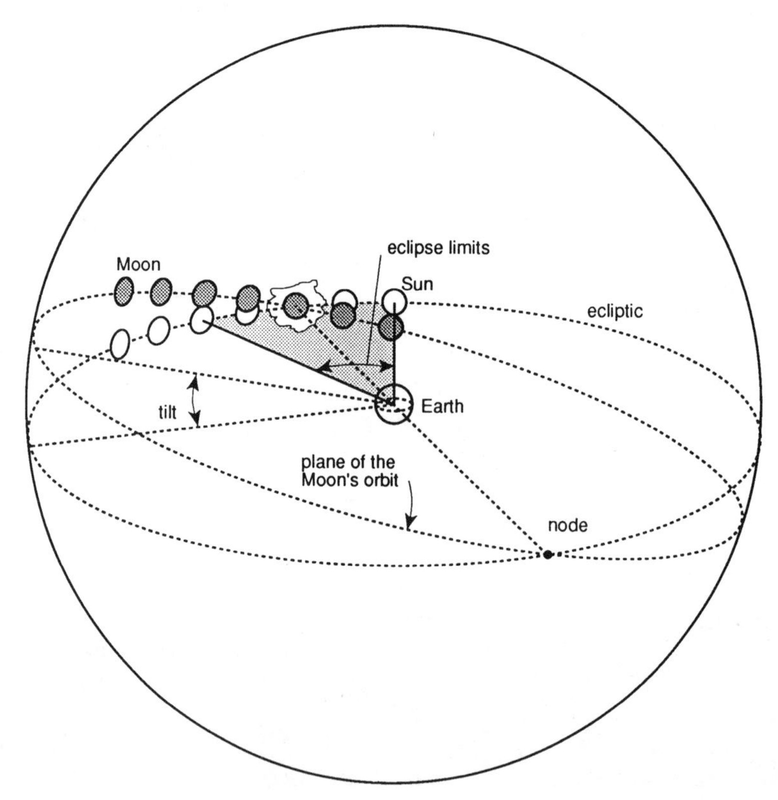

The paths of the Sun and the Moon projected onto the celestial sphere illustrate why eclipses occur only when the Sun is near the intersection (node) of the ecliptic and the path of the Moon. The plane of the Moon's orbit is tilted approximately 5 ° to the ecliptic plane.

A solar eclipse can occur only when the Sun is near one of the nodes as the Moon passes. If the Sun stood motionless in a part of the sky away from the nodes, there would be no eclipses, and you would not be agonizing over this. But the Earth is moving around the Sun, and, as it does so, the Sun appears to move eastward once around the sky, through all the constellations of the zodiac, completing that journey in one year. In that yearly circuit, the Sun must cross the two nodes of the Moon. Think of it as a street intersection at which the Sun does not pause and runs the stop sign every time. It is an accident waiting to happen. When the Sun nears a node, there is the "danger" that the Moon will be coming and—crash!

No. The Moon is 400 times closer to the Earth than the Sun, so the worst—the best—that can happen is that the Moon will pass harmlessly but stunningly right in front of the Sun. The Sun's apparent pathway in the sky is called the *ecliptic* because it is only when the Moon is crossing the ecliptic that eclipses can happen. Thus twice a year, roughly, there is a "danger period," called an *eclipse season*, when the Sun is crossing the region of the nodes and an eclipse is possible.

The Sun comes tootling up to the node traveling about 1 degree a day.[3] The hot-rod Moon, however, is racing around the sky at about 13 degrees a day. Now if the Sun and Moon were just dots in the heavens, they would have to meet precisely at a node for an eclipse to occur. But the disks of the Sun and Moon each take up about half a degree in the sky. And the Earth, almost 8,000 miles (12,800 km) across, provides an extended viewing platform. Therefore, the Sun needs only to be *near* a node for the Moon to sideswipe it, briefly "denting" the top or bottom of the Sun's face. That will happen whenever the Sun is within 15⅓ degrees of a node.

An "eclipse alert" begins when the Sun enters the danger zone 15⅓ degrees west of one of the Moon's nodes and does not end until the Sun escapes 15⅓ degrees east of that node. The Sun must traverse 30⅔ degrees. Traveling at 1 degree a day, the Sun will be in the danger zone for about 31 days. But the Moon completes its circuit, going through all its phases, and catches up to the Sun every 29.53 days. It is not possible for the Sun to pass through the danger zone before the Moon arrives. A solar eclipse *must* occur each time the Sun approaches a node and enters one of these danger zones, about every half year.

In fact, if the Moon nips the Sun at the beginning of a danger

Eclipse Limits ("Danger Zones")

	maximum*	minimum*
A solar eclipse will occur at New Moon if the Sun's angular distance from a node of the Moon is less than	18°31′	15°21′
A central (total or annular) solar eclipse will occur at New Moon if the Sun's angular distance from a node of the Moon is less than	11°50′	9°55′

(*Limits vary due to changes in the apparent angular size and speed of the Sun and Moon because the orbits of the Earth and Moon are elliptical.)

Sun's apparent eastward movement in the star field each day (due to the Earth orbiting the Sun):	about 1°
Moon's synodic period (from New Moon to New Moon: the time it takes to complete its eastward circuit of the star field and catch up with the Sun again):	29.53 days

Significance

The Sun cannot pass a node without at least one solar eclipse occurring, and two are possible. (If two occur, both will be partial, about one month apart.)

The Sun can pass a node without a central solar eclipse (total or annular) occurring.

zone (properly called an *eclipse limit*), the Sun may still have 30 days of travel left within the zone. But the Moon takes only 29.53 days to orbit the Earth and catch up with the Sun again. So it is possible for the Sun to be nipped by the Moon twice during a single node crossing, thereby creating two partial eclipses within a month of one another.

The closer the Sun is to the node when the Moon crosses, the more nearly the Moon will pass over the center of the Sun's face. In fact, if the Sun is within about 10 degrees of the node at the time of the Moon's crossing, a central eclipse will occur somewhere on Earth. Depending on the Moon's distance from the Earth and the Earth's distance from the Sun, this central eclipse will be total or annular.

Nodes on the March

There should be one or two solar eclipses about every six months, whenever the Sun crosses one of the Moon's nodes. Actually, the Sun crosses the ascending node of the Moon, then the descending node, and returns to the ascending node in only 346.62 days—the *eclipse year.* Within this eclipse year there are two *eclipse seasons,* intervals of 30 to 37 days as the Sun approaches, crosses, and departs from a node. All eclipses will fall within eclipse seasons. There can be no eclipses outside this period of time. Because the Sun crosses

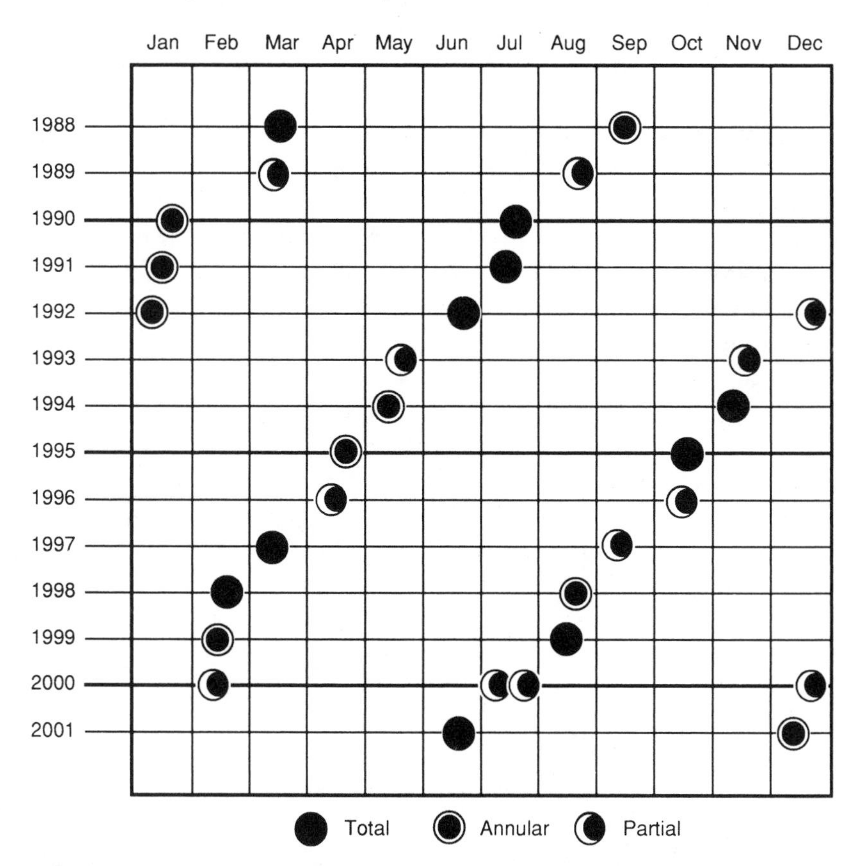

Solar eclipses 1988–2001 plotted to show eclipse seasons. Each calendar year, eclipses occur about 20 days earlier. Consequently, eclipse seasons shift to earlier months in the year. *Drawn by Tim Phelps, after Holger Filling, "Schattenspiele des Mondes,"* Sterne und Weltraum, *Feb. 1990:120*

a node about every 173 days, the eclipse seasons are centered about 173 days apart.

The eclipse year does not correspond to the calendar year of 365.24 days because the nodes have a motion all their own. They are constantly shifting westward along the ecliptic. This regression of the nodes is caused by tidal effects on the Moon's orbit created by the Earth and the Sun. If the nodes did not shift, eclipses would always occur in the same calendar month year after year. If the Sun crossed the nodes in February and August one year to cause eclipses, the eclipses would continue to fall in February and August in succeeding years.

But the eclipse year is 346.62 days, 18.62 days shorter than a calendar year, so each ascending or descending node crossing by the Sun occurs 18.62 days earlier in the calendar year than the previous one of its kind. This migration of the eclipse seasons determines the number of eclipses that may occur each year. One solar eclipse must occur each eclipse season, so there have to be at least two solar eclipses each year (although both may be partial). But because of the width of the eclipse limits—up to 19 days on either side of the Sun's node crossing—and the slowness of the Sun's apparent motion, there can be two solar eclipses at each node passage (both partial). Thus, occasionally there will be four solar eclipses in one calendar year.

There can actually be five. Because the eclipse year lasts 346.62 days, almost 19 days less than a calendar year, if a solar eclipse occurs before or on January 18 (or January 19 in a leap year), that eclipse year could conceivably bring two solar eclipses in January and two more around July. That eclipse year would then end in mid-December, and a new eclipse year would begin in time to provide one final solar eclipse before the end of December.

Heavenly Rhythm

Eclipses, then, are like fresh fruit—available only in season. Ancient peoples who kept written records, such as the Chaldeans from about 747 B.C. on, noticed after decades of observation that eclipses happen only at certain times of the year. These eclipse seasons are separated from one another by either five or six New Moons.

From the expanse of their New Babylonian Empire in the Mid-

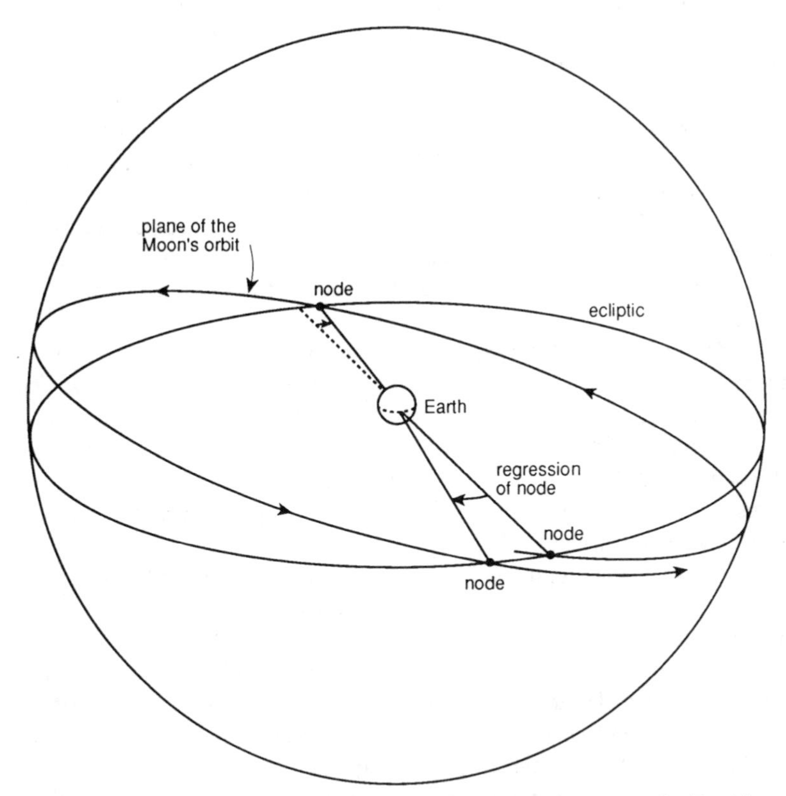

Each time the Moon completes an orbit around the Earth, it crosses the Earth's orbit at a point west of the previous node. Each year the nodes regress 19.4°, making a complete revolution in 18.61 years.

Solar Eclipses Outnumber Lunar Eclipses

There are more solar than lunar eclipses. This realization usually comes as a surprise because most people have seen a lunar eclipse, while relatively few have seen an eclipse of the Sun. The reason for this disparity is simple. When the Moon passes into the shadow of the Earth to create a lunar eclipse, the event is seen wherever the Moon is in view, which includes half the planet. Actually, a lunar eclipse is seen from more than half the planet because during the course of a lunar eclipse (up to 4 hours), the Earth rotates so that the Moon comes into view for additional areas.

In contrast, whenever the Moon passes in front of the Sun, the shadow it creates—a solar eclipse—is small and touches only a tiny portion of the surface of the Earth. On the average, your house will be visited by a total eclipse of the Sun only once in about 410 years.*

To be touched by the dark shadow (umbra) of the Moon is quite rare. But from either side of the path of a total eclipse, stretching northward and southward 2,000 miles (3,200 km) and sometimes more, an observer sees the Sun partially eclipsed. Even so, this band of partial eclipse covers a much smaller fraction of the Earth's surface than a lunar eclipse. So more people have seen lunar eclipses than partial solar eclipses, and only a tiny fraction of people, about one in 10,000, have witnessed a total solar eclipse.

Yet, to the unaided eye, solar eclipses are substantially more frequent than lunar eclipses. Theodor von Oppolzer, in his monumental *Canon of Eclipses,* published shortly after his death in 1887, attempted to compute, with the help of a number of assistants, all eclipses of the Sun and Moon from 1208 B.C. to A.D. 2161. He cataloged about 8,000 solar eclipses and 5,200 lunar eclipses. He found about three solar eclipses for every two lunar eclipses.

This ratio can be misleading, however. Oppolzer counted all solar eclipses, whether they were total (the Moon's *umbra* touches the Earth) or partial (only the Moon's *penumbra* touches the Earth). But for lunar eclipses, Oppolzer counted only those in which the Moon was totally or partially immersed in the Earth's *umbra.* He did not count *penumbral* lunar eclipses because they are virtually unnoticeable. If he had included penumbral lunar eclipses in his census so as to compare the number of all forms of solar and lunar eclipses, the ratio would have been close to even, with solar eclipses barely prevailing.

The reason why solar eclipses slightly outnumber lunar eclipses is most easily visualized if you imagine looking down on the Sun-Earth-Moon system and if you start by considering only total solar and lunar eclipses.

The Moon will be totally eclipsed whenever it passes into the shadow of the Earth—between c and d on the diagram. At the Moon's average

---

dle East, the Chaldeans could see only about half the lunar eclipses and only a small fraction of the solar eclipses, so, for them, eclipse seasons were not periods during which one to three eclipses would necessarily occur. Instead, eclipse seasons were times of "danger" when an eclipse was possible. One of the two great celestial lights might be partially or totally darkened. The Chaldeans did not realize that the

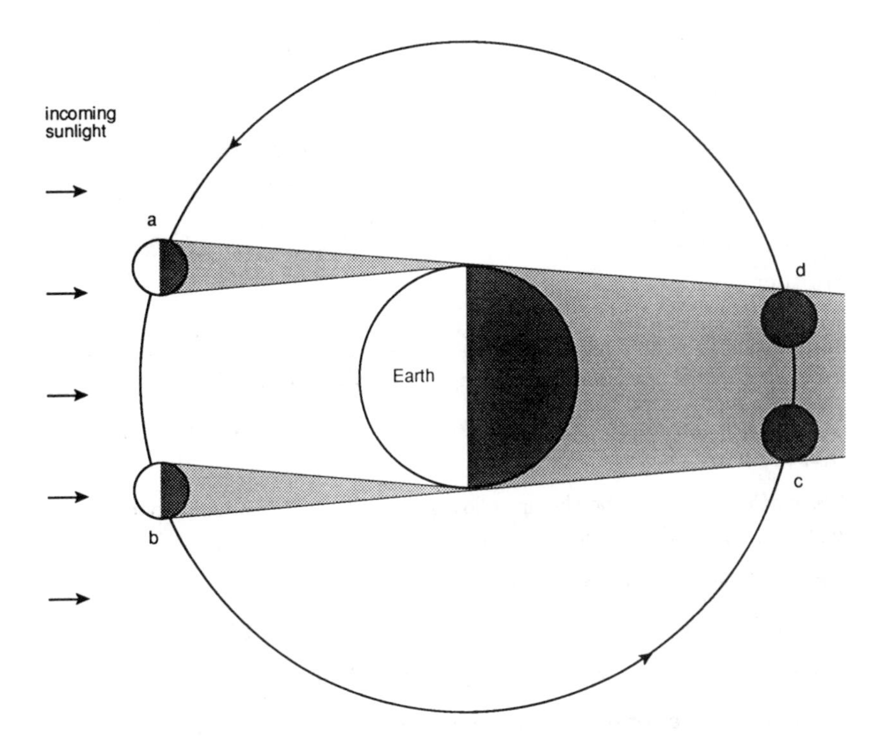

distance from Earth, that shadow is about 2.7 times the Moon's diameter. But there will be a total eclipse of the Sun whenever the Moon passes between the Earth and Sun—between points a and b. The distance between a and b is longer than between c and d, so total solar eclipses must occur slightly more often.

*Owen Gingerich, Charles Kluepfel, and Jean Meeus, Letter to Editor, *Sky & Telescope* 62 (Aug. 1981): 118.

invisibility of an eclipse simply meant that it was occurring somewhere else on Earth.

As time passed and they accumulated more records, the Chaldeans and other ancient peoples recognized that a specific eclipse occurred a precise number of days after a previous eclipse and before a subsequent one. Eclipses had a long-term rhythm of their own.

The most famous and, perhaps, most useful of these eclipse rhythms was the *saros*, discovered by the Chaldeans and inscribed on clay tablets in their cuneiform writing.[4] The Chaldeans noticed that 6,585 days (18 years 11 days) after virtually every lunar eclipse, there was another very similar one. If the first was total, the next was almost always total. And these eclipses, separated by 18 years, occurred in the same part of the sky, as if they were related to one another.

Frequency of Solar Eclipses

Solar eclipses by types: Total 26.9%; Annular 33.2%; Annular/Total 4.8%; Partial 35.2%

Solar eclipses outnumber lunar eclipses almost 3 to 2 (excluding penumbral eclipses, which are seldom detectable visually)

Annular eclipses outnumber total eclipses about 5 to 4

Solar eclipses per century (average over 4,530 years): 237.8

Maximum number of solar eclipses per year: 5 (4 will be partial)

Minimum number of solar eclipses per year: 2 (both can be partial)

Maximum number of *total* solar eclipses per year: 3

Minimum number of *total* solar eclipses per year: 0

Maximum number of solar and lunar eclipses per year: 7 (4 solar and 3 lunar or 5 solar and 2 lunar)

Minimum number of solar and lunar eclipses per year: 2 (2 solar and 0 lunar)

Examples of years in which only 2 solar eclipses occur: 1989, 1990, 1991, 1993

Examples of years in which 5 solar eclipses occur: 1805, 1935, 2206, 2709

Maximum diameter of the Moon's shadow cone (umbra) as it intercepts the Earth to cause a total eclipse: 167 miles (269 km)

Maximum diameter of the Moon's "anti-umbra" as it intercepts the Earth to cause an annular eclipse: 194 miles (313 km)

In a sense, they were. Imagine that a total solar eclipse occurs one day at the Moon's descending node. After 6,585 days, the Moon has completed 223 lunations (synodic periods) of 29.53 days each and returned to New Moon at that same node. In that same period of 6,585 days, the Sun has endured 19 eclipse years of 346.62 days each and has returned to the descending node, forcing another solar eclipse to occur. Because 6,585 days is very close to an even 18 calendar years, this solar eclipse occurs at the same season of the year as its predecessor, and with the Sun close to the same position in the zodiac that it occupied at the eclipse 18 years earlier. Even though 18 years have intervened, these two eclipses certainly seem to be relatives.

All the more so because another lunar cycle crucial to eclipses has a multiple that also adds up to 6,585 days. That cycle is the *anomalistic month*—the time it takes the Moon in its elliptical orbit around the Earth to go from *perigee* (closest to Earth) to *apogee* (farthest) and back to *perigee*.[5] When the Moon is near perigee, its angular size in the sky is just slightly larger, creating eclipses that are total rather than annular. The anomalistic month is 27.55 days long and 239 of these cycles add up to 6,585 days. If the previous eclipse occurred with the Moon at perigee, the new eclipse will occur with the Moon near perigee, providing another total eclipse.

So, after a time interval of 6,585 days, the geometrical configuration of the Sun, Moon, and Earth is nearly repeated. Another solar eclipse occurs at the same node, at the same season, in the same part of the sky. If it was a total eclipse before, it will almost certainly be a total eclipse again. Take the date of any solar or lunar eclipse and add 6,585.32 days to it and you will accurately predict a subsequent eclipse of the same kind that will closely resemble the one 18 years earlier. Take the date of *every* solar and lunar eclipse that occurs and keep on adding 6,585.32 days to it and you will have, with rare exceptions, a reliable list of future eclipses.

Flaws in the Rhythm

But these eclipses 6,585 days apart are not identical twins. The match between unrelated periods—the Moon's cycle of phases, the Sun's eclipse year cycle, and the Moon's cycle from perigee to perigee —is not perfect.

223 synodic periods of the Moon (lunations) at 29.5306 days each	=	6,585.32 days
19 eclipse years of the Sun at 346.6201 days each	=	6,585.78 days
239 anomalistic months of the Moon (revolution from perigee to perigee) at 27.55455 days each	=	6,585.54 days

These synchronizing cycles are out of step with one another by fractions of a day. And those fractions of a day have their consequences.

Consider first that 223 synodic periods of the Moon amount to 6,585.32 days or, in calendar years, 18 years 11⅓ days (18 years 10⅓ days if five leap years intervene). Because the saros period is 18 years 11⅓ days, each subsequent eclipse occurs about one-third of the way around the world westward from the one before it. For a lunar eclipse, visible to half the planet, this westward shift usually does not push the

---

Duration of Totality of Solar Eclipses

Total (minutes:seconds)

Longest possible: 7:31

Longest from 2004 B.C. to 2526 A.D.: July 16, 2186 A.D., 7:29

Longest from 2004 B.C. to present: June 15, 744 B.C., 7:28

Longest in last 2,000 years: June 27, 363 A.D., 7:24

Longest in last 1,000 years: June 9, 1062 A.D., 7:21

Longest in the 20th century: June 8, 1937, 7:4; June 20, 1955, 7:8; June 30, 1973, 7:4. *(All part of same saros series that includes historic eclipses of 1919 and 1991.)*

Number of eclipses with 7 minutes or more of totality in 21st century: 0

Number of eclipses with 7 minutes or more of totality from July 1, 1098 to June 8, 1937 (839 years): 0

Annular (minutes:seconds)

Longest possible: 12:30

Longest from 2004 B.C. to 2526 A.D.: December 7, 150 A.D., 12:23

---

eclipse out of view. For a solar eclipse, however, visible over a narrow swath of the Earth's surface, the subsequent eclipse in that saros series would almost never be visible from the initial site. The eclipse would be happening in an entirely different part of the world.

After three saros cycles—54 years 34 days—the eclipse would be back to its original longitude, but it would have shifted, on the average, about 600 miles (1,000 km) northward or southward, taking totality and even major partiality out of view for the original observer.[6]

A saros period truly does predict with accuracy that a solar eclipse will happen, but it would have been extremely hard for an ancient astronomer to confirm that the predicted eclipse had taken place and thus difficult for the astronomer or those he served to retain confidence in his solar eclipse predictions.

The Evolving Saros

Consider that 223 lunations (synodic periods) amount to 6,585.32 days, while 19 eclipse years of the Sun consume 6,585.78 days. The difference after 18 years 11⅓ days is 0.46 day. The result is that the Sun is not exactly in the same place in relation to the Moon's node. It is 0.477 degree farther west. If the Moon just barely grazed the western part of the Sun before, it will clip a little more the next time. At each return of the saros, the eclipses become larger and larger partials until the Moon is passing across the center of the Sun's disk, yielding total or annular eclipses. Then, with the passing generations, as the Sun is farther west within the eclipse limit, the eclipses return to partials. Finally, after about 1,300 years, the Sun is no longer within the eclipse limits when the Moon, after 223 lunations, arrives. After about 1,300 years of adding 6,585.32 days to the date of an eclipse to predict the next, the prediction fails. No eclipse occurs. That saros has died.[7]

Finally, consider that 223 lunations amount to 6,585.32 days while 239 anomalistic months of the Moon (perigee to perigee) total 6,585.54 days. If the Moon was at perigee and hence largest in angular size during a central eclipse, the eclipse must be total. But with each succeeding eclipse in that saros series, the Moon is a little farther from perigee, until finally the Moon's angular size is too small to cover the Sun completely, and eclipses become annular.

For people long ago, the discovery of the saros and other eclipse cycles probably brought some comfort that eclipses, however dire their interpretation, were part of nature's rhythms. Using these rhythms, it was possible to predict future eclipses without knowing anything about the mechanism that produced them. Even after people understood the causes of eclipses, it was far easier to predict their occurrence by the saros (or other cycles) than to calculate all the factors surrounding the varying motions and apparent sizes of the Sun, Moon, and Earth.

Erratic Intervals between Total Solar Eclipses

Location	Dates	Interval
An average town		average of 410 years
Unusually Long Intervals		
Jerusalem	September 30, 1131 B.C. July 4, 336 B.C.	795 years
London	March 20, 1140 May 3, 1715	575 years
United States mainland	February 26, 1979 August 21, 2017	38 years
Unusually Short Intervals		
Brisbane, Australia (region of)	April 5, 1856 March 25, 1857	11½ months
New Guinea (southern)	June 11, 1983 November 22, 1984	1½ years
Kiev, U.S.S.R. (region of)	March 14, 190 B.C. July 17, 188 B.C. October 19, 183 B.C.	3 in 7½ years
Sumatra	May 18, 1901 January 14, 1926 May 9, 1929	3 in 28 years
Jerusalem (region of)	March 1, 357 B.C. July 4, 336 B.C. April 2, 303 B.C.	3 in 54 years

Today, with modern electronic computers to crunch orbits and tilts and wobbles into accurate eclipse predictions, the saros would seem to be an anachronism. Yet it is interesting to view any single total eclipse as a member of an evolving family that had its origin in the distant past, has gradually risen from insignificance to great prominence, and will inevitably decline into oblivion. Chapter 13 explores the genealogy of one very special eclipse.

Location	Dates	Interval
Scotland	March 7, 1598 April 8, 1652 August 12, 1654	3 in 56 years
Spain	July 8, 1842 December 22, 1870 May 28, 1900 August 30, 1905	4 in 63 years
Yellowstone National Park	July 29, 1878 January 1, 1889	10½ years
Mexico City	March 7, 1970 July 11, 1991	21 years
New York City	June 16, 1806 January 24, 1925	119 years
Lobito, Angola	June 21, 2001 December 4, 2002	1½ years
Confluence of Ohio and Mississippi rivers (southeast Missouri, southern Illinois, and western Kentucky)	August 21, 2017 April 8, 2024	6½ years
Florida panhandle (Pensacola to Tallahassee)	August 12, 2045 March 30, 2052	6½ years

Creating the Ultimate Eclipse

What conditions provide the longest total eclipse of the Sun?

First, the Moon should be near maximum angular size, which means it should be near perigee—the point in its orbit where it is closest to Earth. That happens once every 27.55 days (an anomalistic month).

Second, the Sun should be near minimum angular size, which means the Earth should be near aphelion—the point in its orbit where it is farthest from the Sun. That happens once every 365.26 days (an anomalistic year). At present, aphelion occurs in early July.

Third, to prolong the eclipse as much as possible, the Moon's eclipse shadow must be forced to travel as slowly as possible. During a solar eclipse, the Moon's shadow is moving about 2,100 miles (3,380 km) an hour with respect to the center of the Earth. But the Earth is rotating from west to east, the same direction that the eclipse shadow travels. At a latitude of 40 degrees north or south of the equator, the surface of the Earth turns at about 790 miles (1,270 km) an hour, slowing the shadow's eastward rush by that amount. At the equator, the Earth's surface rotates at 1,040 miles (1,670 km) an hour, slowing the shadow's speed to only 1,060 miles (1,710 km) an hour, thereby prolonging the duration of totality, Baily's Beads, and all phases of the eclipse.

Finally, to prolong the eclipse just a few seconds more, we should be where the surface of the rotating Earth brings us closest to the Moon. This occurs where the Moon is directly overhead and places us about 4,000 miles (6,400 km) closer than when we view the Moon rising or setting. Minimizing the Moon's distance in this way slightly increases its angular size and thereby maximizes its eclipsing power. The latitude where the Moon stands overhead varies with the seasons as the Sun appears to oscillate 23½ degrees north and south of the equator. Thus the peak duration of totality for an exceptionally long eclipse rarely occurs at the equator but *always* occurs within the tropics. The maximum duration of totality for a solar eclipse has been calculated by Isabel M. Lewis to be 7 minutes 31 seconds.

Basic Sun and Moon Data

Diameter of Sun: 864,900 miles (1,392,000 km)

Sun's mean distance from Earth: 93,000,000 miles (149,600,000 km)

Diameter of Moon: 2,160 miles (3,476 km)

Moon's mean distance from Earth: 238,900 miles (384,500 km)

Ratio of Sun's diameter to Moon's diameter: 400.5

Ratio of Sun's mean distance to Moon's mean distance: 389.1

Mean orbital speed of Moon: 2,290 miles (3,680 km) per hour

Speed through space of Moon's shadow during a solar eclipse (not the same as the Moon's orbital speed because of the orbital motion of Earth): 2,100 miles (3,380 km) per hour

Orbital eccentricity of Moon: 0.05490

Inclination of Moon's orbit to plane of Earth's orbit (ecliptic) (varies due to tidal effects of Earth and Sun): 5°08′ (mean); 5°18′ (maximum); 4°59′ (minimum)

Regression (westward drift) of Moon's nodes: 19.4° per year

Period for Moon's nodes to regress all the way around its orbit: 18.61 years

---

I look up. Incredible! It is the eye of God. A perfectly black disk, ringed with bright spiky streamers that stretch out in all directions.

Jack B. Zirker (1984)

3

A Quest to Understand

Ancient peoples around the world have left monuments and symbols of their reverence for the sky and the results of their efforts to record celestial motions. More than 2,500 years ago, some people could predict or at least warn of the possibility of eclipses, especially lunar eclipses, the fading of the Moon as it passed into the Earth's shadow.

Stonehenge

The earliest and most famous of monuments that testify to a people's high level of astronomical knowledge is Stonehenge, near Salisbury in southern England. Awesome, haunting, strangely beautiful, Stonehenge is a permanent record of celestial knowledge in stone. Work was in progress on Stonehenge a generation before construction began on the first pyramids in Egypt. Stonehenge was already abandoned, a desolate mystery, when Moses led the Exodus from Egypt.

The familiar silhouette of archways, now mostly fallen, is the last of four major stages of development at Stonehenge that ceased about 1500 B.C. The stones formed a circle of 30 linked archways, approximating the days in a lunar month.[1] Inside this Sarsen Circle was a horseshoe of five even larger freestanding archways, the tri-

Aerial view of Stonehenge. *English Heritage*

lithons, with uprights that weigh up to 50 tons. These massive, shaped boulders were dragged from a quarry 20 miles (32 km) away to codify in stone the discoveries of an earlier people.

The most famous feature of Stonehenge is the line of sight from the center of the monument toward the northeast through an archway in the Sarsen Circle and over the Heel Stone, a 35-ton boulder set upright 245 feet (75 m) away. From this position, an observer sees the approximate point on the horizon where the Sun rises on the first day of summer, when it is farthest north of the equator and daytime lasts longest. With this alignment and others, the users of Stonehenge could time the beginning of summer and winter with high precision to create an accurate solar calendar, which could have been of great benefit to farmers. Yet these massive stone archways, raised in the last phase of construction at Stonehenge, show little new in their orientations beyond what archaeologists have found in the original and essential Stonehenge, begun 1,300 years earlier.

Stonehenge was begun about 2800 B.C. by a people who had no written language, no wheeled vehicles, no draft animals, and no metal tools. To dig holes in the ground, they used the antlers of deer.

The initial Stonehenge consisted of a circular embankment 350 feet (107 m) in diameter, four marker stones set in a rectangle, some

Viewed from the center of Stonehenge, the Sun rises just to the left of the remaining Heel Stone at the beginning of summer, the longest day of the year. *English Heritage*

postholes, and the Heel Stone.[2] The Heel Stone was apparently the first of the great boulders brought to the site as construction commenced. But it may not have stood alone. A similar huge stone stood just to its left as seen from the center of Stonehenge.[3] In that ancient time, the Sun at the beginning of summer rose between the famed Heel Stone and its now-vanished companion, and the alignment with sunrise at the summer solstice was probably exact.

For someone standing at the center of Stonehenge, the embankment served to level the horizon of rolling hills. Within the embankment, four stones—the Station Stones—outlined a rectangle within the circle. The sides of this rectangle offered interesting lines of sight. The short side of the rectangle pointed toward the same spot on the horizon that the two Heel Stones framed, the position where the Sun rose farthest north of east, marking the commencement of summer. Facing in the opposite direction along the short side of the rectangle, an observer would see the place where the Sun set farthest south of west, signaling the beginning of winter.

In contrast, the long sides of the rectangle provided alignments for crucial rising and setting positions of the Moon. Looking southeast along the length of the rectangle, an observer was facing the point on the horizon where the summer Full Moon would rise farthest south. In the opposite direction, looking northwest, an early astronomer's gaze was led to the spot on the horizon where the winter Full Moon would set farthest north. These positions marked the north and south limits of the Moon's motion.

The structure of Stonehenge offers additional testimony to its builders' efforts to understand the motion of the Moon. Evidence of small holes near the remaining Heel Stone strongly suggests that the users of Stonehenge observed and marked the excursion of the Moon as much as 5 degrees north and south of the Sun's limit.[4] This motion above and below the Sun's position is caused by the tilt of the Moon's orbit to the Earth's path around the Sun. Because of this tilt, the Moon does not pass directly in front of the Sun (a solar eclipse) or directly into the Earth's shadow (a lunar eclipse) each month.

Because the builders of Stonehenge had discovered and accurately recorded the range in the rising and setting positions of the Sun and Moon and had built a monument that marked these positions with precision, they may have been able to recognize when the Moon was on course to intercept the position of the Sun, to cause a solar eclipse. Perhaps they could tell when the Moon was headed for a position directly opposite the Sun, which would carry it into the shadow of the Earth for a lunar eclipse. They almost certainly could not predict where or what kind of solar eclipse would be seen, but they might have been able to warn that on a particular day or night, an eclipse of the Sun or Moon was *possible*.

In the last phase of building at Stonehenge, two concentric circles of holes were dug just outside the Sarsen Circle: one with 30 holes

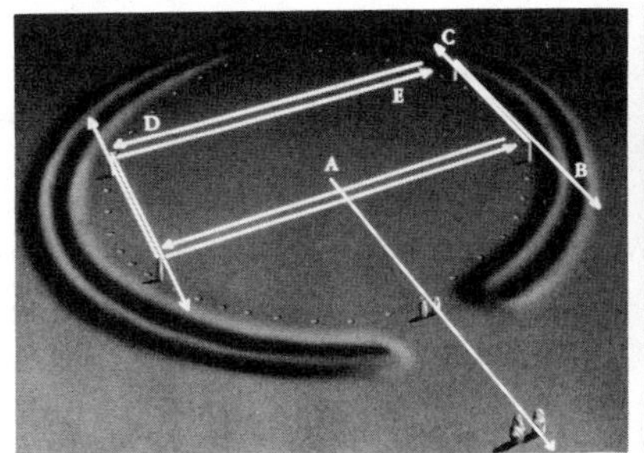

Stonehenge: *left,* first phase of construction showing alignments A and B, northernmost sunrise: first day of summer; C, southernmost sunset: first day of winter; D, southernmost moonrise; E, northernmost moonset; *right,* final phase of construction. *Drawings © 1983 Hansen Plantarium/Salt Lake County*

and the other with 29. These circles reinforce the evidence that astronomers at Stonehenge were counting off the 29½-day cycle of lunar phases, from New Moon to Full Moon and back to New Moon. Eclipses of the Sun can take place only at New Moon; lunar eclipses can occur only at Full Moon. If indeed the lunar phasing cycle was watched carefully, perhaps some ancient genius noticed a periodicity in eclipses as well. With a knowledge of that period, that early astronomer could have converted a mere warning of a possible eclipse into a prediction of a likely eclipse, especially for lunar eclipses, which are visible over half the Earth.[5]

The builders of Stonehenge left no written records of their objectives or results, so we must judge from the monument and its alignments what they knew. Whatever that was, they thought it so worth celebrating that the rulers and apparently the common people were willing to devote vast amounts of time, physical effort, and ingenuity to raising a lasting monument of great size, precision, and beauty.

China

A frequently recounted Chinese story tells that Hsi and Ho, the court astronomers, got drunk and neglected their duties so that they failed to predict (or react to) an eclipse of the Sun. For this, the emperor had them executed. So much for negligent astronomers.

If this story were an account of an actual event, the dynasty mentioned would place the eclipse somewhere between 2159 and 1948 B.C., making it by far the oldest solar eclipse recorded in history. But all serious attempts to identify one particular eclipse as the source of the story have been abandoned as scholars have recognized that the episode is mythological.

In ancient Chinese literature, Hsi-Ho is not two persons but a single mythological being who is sometimes the mother of the Sun and at other times the chariot driver for the Sun. Later, in the *Shu Ching* [Historical Classic], parts of which may date from as early as the seventh or sixth century B.C., this single character is split, not into two, but into six. In the *Shu Ching* story, the legendary Chinese emperor Yao commissions the eldest of the Hsi and Ho brothers "to calculate and delineate the sun, moon, the stars, and the zodiacal markers; and so to deliver respectfully the seasons to the people."[6] In further orders, he sends a younger Hsi brother to the east and another to the south; he orders a younger Ho brother to the west and

The Hsi and Ho brothers receive their orders from Emperor Yao to organize the calendar

another to the north. Each is responsible for a portion of the rhythms of the days and seasons, to turn the Sun back at the solstices and to keep it moving at the equinoxes.

These mythological magicians are always charged with the prevention of eclipses, hence the story that appears later in the *Shu Ching* about the emperor's anger with his servants for failing to *prevent* an eclipse, not just predict or respond ceremonially to it. The story appears in a chapter that is an exhortation by the Prince of Yin, commander of the armies, to government officials to fulfill their duties to the administration, thereby making the emperor "entirely intelligent." If anyone neglects this requirement, "the country has regular punishments for you."

> Now here are Hsi and Ho. They have entirely subverted their virtue, and are sunk and lost in wine. They have violated the duties of their office, and left their posts. They have been the first to allow the regulations of heaven to get into disorder, putting far from them their proper

> business. On the first day of the last month of autumn, the sun and moon did not meet harmoniously in Fang. The blind musicians beat their drums; the inferior officers and common people bustled and ran about. Hsi and Ho, however, as if they were mere personators of the dead in their offices, heard nothing and knew nothing;—so stupidly went they astray from their duty in the matter of the heavenly appearances, and rendering themselves liable to the death appointed by the former kings. The statutes of government say, "When they anticipate the time, let them be put to death without mercy; when they are behind the time, let them be put to death without mercy."[7]

We never hear whether Hsi and Ho were tracked down and executed.

The story of Hsi and Ho as drunken astronomers was a myth. But the myth did come true in a sense about 33 centuries later. Chinese history records that in A.D. 1202, for the second time in four years, the chief court astronomer made an eclipse forecast that was not as accurate as predictions from people with no official scientific credentials or status. "The astronomical officials were found guilty of negligence and severely punished."[8]

The earliest Chinese word for eclipse, *shih*, means "to eat" and referred to the gradual disappearance of the Sun or Moon as if it were eaten by a celestial dragon.[9] The Chinese were early in recording eclipses but late in recognizing their cause. Not until the third or fourth century A.D. did they understand solar and lunar eclipses well enough to be able to predict them accurately.

The New World

In the New World, there were ancient people who, like the Chaldeans and the Chinese, used writing to record eclipses and from these records detected a rhythm by which they could predict or at least warn of the likelihood of eclipses. Those people were the Maya and we know of their achievement through one of their books—one of only four that survived the Spanish conquest and its zealous destruction of the religious beliefs of the native peoples.

All that we know of Maya accomplishments in recognizing the patterns of eclipses comes from the Dresden Codex, written in hieroglyphs and pictures in color paints on (processed) tree bark with pages that open and shut in accordion folds. The book dates from the eleventh century A.D. and is probably a copy of an older work.

We can only wonder what was lost when the conquering Spaniards destroyed by the thousands the books of the Maya and other Mesoamerican peoples. What remains is impressive enough. The Maya realized that discernible eclipses occur at intervals of five or six lunar months. Five or six Full Moons after a lunar eclipse, there was the *possibility* of another lunar eclipse. Five or six New Moons after a solar eclipse, another solar eclipse was *possible*.

The Maya had discovered in practical, observable terms the approximate length of the eclipse year, 346.62 days, and the eclipse half year, 173.31 days. The interval for one complete set of lunar phases is 29.53 days. Six lunations amount to approximately 177.18 days, close

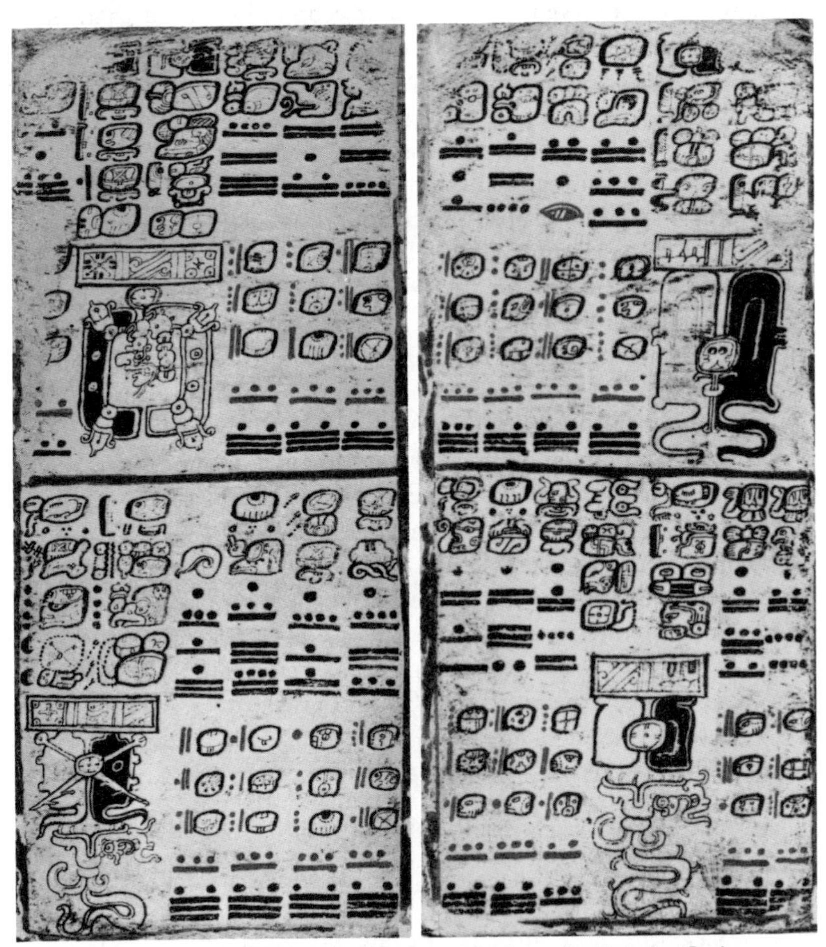

Portion of the Maya solar eclipse prediction tables from the Dresden Codex: at the bottom are the day counts that lead up to a solar eclipse, indicated, *bottom right,* by serpent swallowing a symbol for the Sun. *American Philosophical Society*

enough to the eclipse half year so that there is the "danger" of an eclipse at every sixth New or Full Moon, but not a certainty. After another six lunar months, the passing days have amounted to 354.36, nearly 8 days *too long* to coincide with the Sun's passage by the Moon's node. An eclipse is less likely. As the error mounts, the need increases to substitute a five-lunar-month cycle into the prediction system rather than the standard six-lunar-month count.

Some great genius must have noticed after recording a sizeable number of eclipses that major eclipses were occurring only at intervals of 177 days (six lunar months) or 148 days (five lunar months). Using the date of an observed solar or lunar eclipse, he would then have been able to predict the likelihood of another eclipse, even though in some cases an eclipse would not occur and in others it would not be visible from Mesoamerica.

In the Dresden Codex there are eight pages with a variety of pictures representing an eclipse. Each depiction is different, but most show the glyph for the Sun against a background half white and half black. In two of the pictures, the Sun and background are being swallowed by a serpent. Leading up to each picture is a sequence of numbers: a series of 177s ending with a 148. Each sequence adds up to the number of days in well-known three- to five-year eclipse cycles. At the end of each burst of numbers stands the giant, haunting symbol of an eclipse.

Astronomer-anthropologist Anthony F. Aveni notes that "The reduction of a complex cosmic cycle to a pair of numbers was a feat equivalent to those of Newton or Einstein and for its time must have represented a great triumph over the forces of nature."[10]

From the Maya, we have the numbers that demonstrate one of the greatest of their many discoveries about the rhythms of the sky, but we have no account of the emotion the astronomer-priests or the common folk felt when they observed an eclipse. Perhaps the closest we can come is a passage in the Florentine Codex of the Aztecs, who inherited and used the Mesoamerican calendar but apparently knew little of the astronomy discovered by the Maya a thousand years and more before.

> When the people see this, they then raise a tumult. And a great fear taketh them, and then the women weep aloud. And the men cry out, [at the same time] striking their mouths with [the palms of] their hands. And everywhere great shouts and cries and howls were raised. . . . And they said: "If the sun becometh completely eclipsed, nevermore will he give light; eternal darkness will fall, and the demons will come down. They will come to eat us!"[11]

[T]here was at the same time something in its singular and wonderful appearance that was appaling: and I can readily imagine that uncivilised nations may occasionally have become alarmed and terrified at such an object . . . .

Francis Baily (1842)

4

Eclipses in Mythology

Two great lights brighten the heavens. Life depends on them. The disappearance of one threatens the order of the universe and life itself. Through the ages, most cultures responded to eclipses of the Sun and the Moon with folklore to explain the eerie events.

Solar eclipse mythology might be divided into several themes, and each of these themes is found scattered throughout the world:

A celestial being (usually a monster) attempts to destroy the Sun.
The Sun fights with its lover the Moon.
The Sun and Moon make love and discreetly hide themselves in darkness.
The Sun god grows angry, sad, sick, or neglectful.[1]

Within these myths is a great truth. The harmony and well-being of Earth are dependent on the Sun and the Moon. Abstract science cannot convey this profound realization as powerfully as a myth in which the celestial bodies come to life.

The Sun for Lunch

Most often in mythology a solar eclipse is considered to be a battle between the Sun and the spirits of darkness. The fate of Earth and its inhabitants hangs in the balance. With so much at stake, the people enduring an eclipse were anxious to help the Sun in this struggle if they could.

In the mythology of the Norse tribes, Loki, an evil enchanter, is put in chains by the gods. In revenge, he creates giants in the shape of wolves. The mightiest is Mânagarmer (Moon-Wolf, also called Hati), who causes a lunar eclipse by swallowing the Moon. Sköll, another of these wolflike giants, follows the Sun, always seeking a chance to devour it. Old French and German expressions and incantations echo this belief: "God protect the Moon from wolves."

In India, Rāhu is one of the Asurases, demons who are the elder brothers of the gods. The Asurases and the gods fought over the possession of Lakṣmī, goddess of wealth and beauty, and the possession of ambrosia. The Asurases had captured the ambrosia and Rāhu was drinking it when the god Nārāyana caught up with him, threw his discus, and sliced Rāhu in half. That is why Rāhu has a head but no body. The head flew off into the sky where it attacks the Sun and Moon, swallowing them to cause an eclipse. The severed lower body of Rāhu is Ketu and has become the constellations.

Buddhism placed the Buddha in a position to correct the celestial terrorism of Rāhu. In the midst of an eclipse, the Sun or Moon cries out to the Buddha for help. "Rāhu," says the Buddha, "let go of the [Sun], for the Buddhas pity the world." Rāhu departs in terror, fearing that, if he harms the Sun or the Moon, the Buddha will cause his head to shatter into seven parts.[2]

The Indian tale of Rāhu spread eastward into China and northward into Mongolia and eastern Siberia, where the monster's name became Arakho or Alkha. The Buryat people, living east of Lake Baikal, told of Alkha, a monster who continually pursued and swallowed the Sun and Moon. Finally, the gods were exasperated by the repeated darkening of the world and cut Alkha in two. His lower body fell to Earth, but the upper portion lives on and continues to haunt the sky. This is why the Sun and Moon still disappear from time to time: Alkha swallows them. But they soon reappear because Alkha's body cannot retain them. When eclipses are in progress, say the Buryats, the Sun and Moon pray for help, and the people respond by screaming and by throwing stones and shooting weapons into the sky to scare away the monster.

Another Buryat myth says that the eclipse maker is Arakho, a beast who formerly lived on Earth. In those days long ago, the people were quite hairy. Arakho roamed the Earth eating the hair off their bodies until the people had become the nearly hairless creatures they are today. This annoyed the gods, who chopped Arakho in two. Thus Arakho no longer grazes on human hair; instead the upper portion of his body, which is still alive, eats the Sun and Moon, causing eclipses. Rāhu appears again in Indonesia and Polynesia as Kala Rau, all head and no body, who eats the Sun, burns his tongue, and spits out the Sun.

Ancient Egypt produced the tale of a black pig, the evil god Set in disguise, who leaped into the eye of Horus, the Sun god. The story becomes confused before we learn how the eye was healed, but it may have been the work of Thout, the Moon, who regulates such disturbances as eclipses and is also the healer of eyes.[3]

In northwestern France (Ille-et-Vilaine), the people recognized that eclipses occur when the Moon blocks the Sun from view. For

Egyptian emblem of the winged Sun on the Gateway of Ptolemy at Karnak

Drawing of corona by Samuel P. Langley from the top of Pike's Peak (July 29, 1878). The elongated corona resembles the Egyptian emblem of the winged Sun.

them, the Moon's aggression took a different form. If the Moon were to cover the Sun entirely, it would stick in that position so that the Sun would never shine again.[4]

From around the world come stories of many different monsters intent on devouring the Sun and Moon. In China, it was a heavenly dog who causes eclipses by eating the Sun. In South America, the Mataguaya Indians of the pampas saw eclipses as a great bird with wings outspread, assailing the Sun or Moon. In Armenian mythology, eclipses were the work of dragons who sought to swallow the Sun and Moon. By contrast, another Armenian myth says that a sorcerer can stop the Sun or the Moon in their courses, deprive them of light, and even force them down from the skies. Despite the Moon's size, once it has been brought down to Earth, the sorcerer can milk it like a cow.[5]

On rare occasions in mythology, the Sun and Moon are not totally innocent victims. In a variant Hindu myth, the Sun and Moon borrowed money from a member of the savage Ḍom tribe and failed to pay it back. In retribution, the Ḍom occasionally devours the two heavenly bodies.[6]

The Original Black Holes

In two instances, the hungry monster who swallows the Sun or the Moon becomes quite scientifically sophisticated in character. A western Armenian myth, said to be borrowed from the Persians, tells of two dark bodies, the children of a primeval ox. These dark bodies orbit the Earth closer than the Sun and Moon. Occasionally they pass in front of the Sun or Moon and thereby cause an eclipse.[7]

Still more remarkable is a Hindu myth that speaks of the Navagrahas, "the nine seizers." These nine "planets" that wander through the star field include the usual seven familiar to the Greeks—Sun, Moon, Mercury, Venus, Mars, Jupiter, and Saturn—plus Rāhu and Ketu, "regarded as the ascending and descending nodes" of the Moon, the shifting points in the sky where the Moon crosses the apparent path of the Sun.[8] Thus, quite correctly, the Sun would be at risk of an eclipse whenever it passed Rāhu or Ketu.

Buddhism carried Rāhu and Ketu from India to China in the first century A.D., where they became Lo-Hou and Chi-Tu. They were imagined as two invisible planets positioned at the nodes in the Moon's path: Lo-Hou at the ascending node and Chi-Tu at the descending node. These "dark stars" were numbered among the planets and were considered to be the cause of eclipses.[9]

Love, Marriage, and Domestic Violence

A Germanic myth explained eclipses differently. The male Moon married the female Sun. But the cold Moon could not satisfy the passion of his fiery bride. He wanted to go to sleep instead. The Sun and Moon made a bet: whoever awoke first would rule the day. The Moon promptly fell asleep, but the Sun, still irritated, awoke at 2 A.M. and lit up the world. The day was hers; the Moon received the night. The Sun swore she would never spend the night with the Moon again, but she was soon sorry. And the Moon was irresistibly drawn to his bride. When the two come together, there is a solar eclipse, but only briefly. The Sun and Moon begin to reproach one another and fall to quarreling. Soon they go their separate ways, the Sun blood red with anger.

It was not always a fight between the Sun and Moon that caused an eclipse. Sometimes it was love, and modesty. The Tlingit Indians of the Pacific coast in northern Canada explained a solar eclipse as the Moon-wife's visit to her husband. Across the continent, in southeastern Canada, the Algonquin Indians also envisioned the Sun and Moon as loving husband and wife. If the Sun is eclipsed, it is because he has taken his child into his arms.[10]

For the Tahitians, the Sun and Moon were lovers whose union creates an eclipse. In that darkness, they lose their way and create the stars to light their return.

An Angry Sun

Sometimes, as in a folktale from eastern Transylvania, it is the perversion of mankind that brings on an eclipse. The Sun shudders, turns away in disgust, and covers herself with darkness. Stinking fogs gather. Ghosts appear. Dogs bark strangely and owls scream. Poisonous dews fall from the skies, a danger to man and beast. Neither humankind nor animals should consume water or eat fresh fruit or vegetables. Such beliefs persisted into the nineteenth century. The poisonous dew that supposedly accompanied eclipses could be the source of an outbreak of the plague or other epidemics. If people had to leave their homes, they wrapped a towel around their mouths and noses to strain out the noxious vapors. Clothes caught drying outdoors during a solar eclipse were considered to be infected.

The Germans were not alone in their belief that a solar eclipse brought a dangerous form of precipitation. Eskimos in southwestern Alaska believed that an unclean essence descended to Earth during an eclipse. If it settled on utensils, it would produce sickness. Therefore, when an eclipse began, every Eskimo woman turned all her pots, buckets, and dishes upside down.[11]

Yet it was not always an angry Sun that brought darkness to the Earth. When U.S. Coast Survey scientist George Davidson observed the eclipse of August 7, 1869, from Kohklux, Alaska, he found that the Indians there attributed the eclipse to an illness of the Sun. He had alerted them to the impending eclipse, but they doubted it. Halfway through the partial phase, the Indians and their chief quit work and hid in their houses: "[T]hey looked upon me as the cause of the Sun's being 'very sick and going to bed.' They were thoroughly alarmed, and overwhelmed with an indefinable dread."[12]

Sometimes in mythology, an eclipse is not a monster devouring the Sun, not a sickness of the Sun, not a fight between the Sun and Moon, not even the result of the always abundant sins of mankind. Sometimes an eclipse is what in sports would be called an unforced error. The Bella Coola Indians of the Pacific coast in Canada had a myth that began with a remarkable observational description of the Sun's apparent annual pathway though the sky. The trail of the Sun, they said, is a bridge whose width is the distance between the summer and winter solstices, the northernmost and southernmost positions of the Sun. In the summer, the Sun walks on the right side of this bridge; in the winter, he walks on the left. The solstices are where the Sun sits down. Accompanying the Sun on his journey are three guardians who dance about him. Sometimes the Sun simply drops his torch, and thus an eclipse occurs.[13]

Warding Off Evil

Corruption and death are a frequent theme of eclipse myths. Evil spirits descend to Earth or emerge from underground during eclipses.

On June 16, 1406, says an enlightened and bemused French chronicler, "between 6 and 7 A.M., there was a truly wonderful eclipse of the Sun which lasted nearly half an hour. It was a great shame to see the people withdrawing to the churches and believing that the world was bound to end. However the event took place, and afterward the astronomers gathered and announced that the occurrence was very strange and portended great evil."[14] (Half an hour is too long for totality but the right length for conspicuously reduced light surrounding totality.)

For Hindus, the place to be during an eclipse was in the water, especially in the purifying current of the Ganges.[15] This Hindu practice of immersing oneself in water was known to the French philosopher and popularizer of science Bernard Le Bovier de Fontenelle, as recorded in his *Entriens sur la pluralité des mondes* (Conversations on the Plurality of Worlds) in 1686.

> All over India, they believe that when the sun and moon are eclipsed, the cause is a certain dragon with very black claws which tries to seize those two bodies, wherefore at such times the rivers are seen covered

with human heads, the people immersing themselves up to the neck, which they regard as a most devout position, and implore the sun and moon to defend themselves well against the dragon.[16]

A Participatory Event

From around the world come reports of people trying to assist the Sun and Moon against the peril of eclipse. Screaming, crying, and shouting are supposed to encourage the Sun and Moon to escape the clutches of the evil spirit. Historians recorded that Germans watching a lunar eclipse in the Middle Ages chanted in unison, "Win, Moon." The Sun and Moon were the supreme gods of the Indians of Colombia, in South America. When these gods were threatened by an eclipse, the people seized their weapons and made warlike sounds on their musical instruments. They also shouted to the gods, promising to mend their ways and work hard. To prove it, they watered their corn and worked furiously with their tools during the eclipse.[17]

People frequently augmented their voices with the clanging of metal pots, pans, and knives. The Chippewa Indians in the northeastern United States and southeastern Canada went even further. Seeing the Sun's light being extinguished, they shot flaming arrows into the sky hoping to rekindle the Sun.[18] The Sencis of eastern Peru did the same, but to scare off a savage beast attacking the Sun.

Ethnographers descended on the Kalina tribe in Surinam to collect their folklore and watch their behavior as the total eclipse of June 30, 1973, neared. In Kalina mythology, the Sun and Moon are brothers. They usually get along well together, but occasionally they have sudden and ferocious quarrels that endanger mankind. At such times, it is important to separate the combatants by making a maximum of noise: banging on tools, hollow objects, and instruments. The fading of the Sun or Moon means that one has been knocked unconscious. When this happens, the tribesmen yell, "Wake up, Papa!" Papa, here, is a term of respect, not an indication that the Kalina consider themselves descended from the Sun or Moon. After the 1973 eclipse, invisible because of clouds but noticeable because of darkening, the old women (the pot makers) rounded up the children and used branches to spread white clay all over them, somewhat too vigorously for the children's enjoyment. Then they smeared the women and finally the men from head to foot with white clay. The white clay was the blood of the injured Moon that had dripped onto the ground. It was necessary to wash oneself with the Moon's blood to restore purity in man and whiteness to the Moon. After an hour, the tribesmen washed themselves off in the river.[19]

In ancient Mexico and Central America, the most important god was represented as a plumed serpent. For the Maya, he was Kukulcán; for the Aztecs, Quetzalcóatl. At an eclipse of the Moon and most especially during an eclipse of the Sun, a special snake was killed and eaten.[20]

In prehistoric times, screams, cries, banging noises, and prayer may not have been deemed adequate to ward off eclipses or their effects. In many places around the world, human sacrifice was performed at the appearance of unexpected and confusing sights, such as an eclipse or a comet. Yet few eclipse myths refer to human sacrifice, suggesting that this practice had largely been abandoned before most eclipse myths were preserved. An exception was in Mexico and Central America, where the Spanish invaders saw the Aztecs and their neighbors carry out human sacrifice in the early sixteenth century. For the Aztecs, almost any natural or political event was commemorated with a sacrifice. On the occasion of an eclipse, the Sun was in need of help from people, just as he had constant help from the dog Xolotl (Sho-LOT-uhl). Xolotl was the god of human monstrosities (including twins), so it was humpbacks and dwarfs that were sacrificed to the Sun to help him prevail.[21]

Solar eclipses boded ill for everyone, it seems, except prospectors. In Bohemia, people believed that a solar eclipse would help them find gold.

[T]he Sun . . .
In dim eclipse disastrous twilight sheds
On half the nations, and with fear of change
Perplexes monarchs.

John Milton (1667)

5

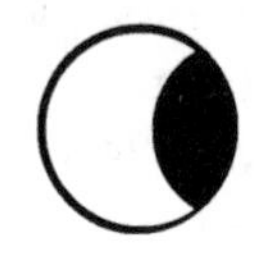

Strange Behavior of Man and Beast

The Human Response

One of the most dramatic responses in history to a total solar eclipse is presented by Herodotus, the first Greek historian, writing around 430 B.C.

> [W]ar broke out between the Lydians and the Medes [major powers in Asia Minor], and continued for five years, with various success. In the course of it the Medes gained many victories over the Lydians, and the Lydians also gained many victories over the Medes. . . . As, however, the balance had not inclined in favour of either nation, another combat took place in the sixth year, in the course of which, just as the battle was growing warm, day was on a sudden changed into night. This event had been foretold by Thales, the Milesian, who forewarned the Ionians of it, fixing for it the very year in which it actually took place. The Medes and Lydians, when they observed the change, ceased fighting, and were alike anxious to have terms of peace agreed upon.

A treaty was quickly made and sealed by the marriage of the daughter of the Lydian king to the son of the Median king.[1] This story indicates the awe that ancient people felt when confronted with a total eclipse of the Sun.

Modern astronomers, armed with the dates of the kings de-

During a battle between the Lydians and the Medes on May 28, 585 B.C., a total eclipse of the Sun occurred. It scared the soldiers so badly that they stopped fighting and signed a treaty.

scribed in the account and a knowledge of the dates and paths of ancient eclipses, have generally settled upon May 28, 585 B.C. as the eclipse to which the story refers, if Herodotus, given to fanciful embellishment, can be trusted about an event that occurred a century before he was born.[2]

In his account, Herodotus credits Thales with predicting the eclipse. If so, Thales would have been the first person *known* to have calculated a future solar eclipse. Cuneiform writing on clay tablets from the Chaldean (or New Babylonian) Empire dating about two centuries later shows recognition of an 18-year-11-day rhythm in eclipses—the saros. Perhaps Thales borrowed this eclipse rhythm from the Babylonians as it was being developed. However, such a rhythm predicts not just the year but the month and precise day of the eclipse. Yet Herodotus seems amazed that Thales could be accurate to "the very year in which it actually took place." Was Herodotus so surprised that Thales could predict an eclipse accurate to the day that he simply could not believe that degree of precision and used the more conservative "year" instead? That would be out of character for the flamboyant Herodotus. Yet predicting a solar eclipse accurate to a year is not much of a trick since there are a minimum of two solar eclipses each year. The problem is to predict a total eclipse for a particular location on Earth. Could Thales have accomplished this? It is doubtful.

The saros period is actually 18 years 11⅓ days, so the Earth has spun through an extra eight hours and the midpoint of the subsequent eclipse falls about one-third of the way around the world westward from the one before it. Thus successive eclipses in a saros series are almost never visible from the same site.

The eclipse in the same saros series that preceded 585 B.C. occurred on May 18, 603 B.C., with an early morning path from the northern portion of the Red Sea to the northern tip of the Persian Gulf, about 600 miles (1,000 km) distant from the end of the path of the May 28, 585 B.C. eclipse. Thales could have heard reports of the 603 B.C. eclipse and used it to calculate the date for the 585 B.C. eclipse. But the saros projection would not have told him where the

---

Tenskwatawa Uses an Eclipse

Tenskwatawa, the Shawnee Prophet (1775?–1837?), was an important Indian religious leader in Ohio and Indiana in the early nineteenth century. He saw great danger for his people as they increasingly adopted the customs of the European settlers, especially alcohol. He urged them to return to traditional Indian ways and to unite into a single Indian nation under the leadership of his brother Tecumseh to resist the encroachment of white men and their fraudulent treaties.

General William Henry Harrison, later president of the United States, was at that time the governor of Indiana Territory, where the Shawnee Prophet was successfully recruiting converts to his Indian religious revival. Seeking to undermine the credibility of the Shawnee Prophet as a shaman, Harrison urged Indians to demand proof from the Prophet that he could perform miracles. Thinking in biblical terms, Harrison asked if Tenskwatawa could "cause the Sun to stand still, the Moon to alter its course, the rivers to cease to flow, or the dead to rise from their graves."

The followers of the Shawnee Prophet did not need such displays, but Tenskwatawa was a canny politician. He proclaimed that on July 16,

---

eclipse would be visible. Thales, then, first of the great Greek philosophers, could have warned of the *possibility* of a solar eclipse, but he could not predict from the saros period that it would be visible in Asia Minor. And there is no evidence that he had the celestial knowledge or the mathematics to calculate it from orbital considerations.

Of course the key to appreciating the story of the solar eclipse that stopped a war is the realization that people long ago were stunned by a total eclipse of the Sun and incredulous that someone could predict such an event. Quite often in ancient history, eclipses are reported to have played a decisive role in the turn of events.

Herodotus tells of another turning point in world history that he says hinged on a solar eclipse. Xerxes and his Persian army were about to march from Sardis to Abydos on their advance toward Greece.

> At the moment of departure, the sun suddenly quitted his seat in the heavens, and disappeared, though there were no clouds in sight, but the sky was clear and serene. Day was thus turned into night; whereupon Xerxes, who saw and remarked the prodigy, was seized with alarm, and sending at once for the Magians, inquired of them the

---

1806, he would blot the Sun from the sky as a sign of his divine powers. Whether he knew of this total eclipse from a British agent or from an almanac is uncertain, but a great many Indians gathered at the Shawnee Prophet's camp as the appointed day dawned clear.

At the proper moment, the Prophet, in full ceremonial regalia, pointed his finger at the Sun, and the eclipse began. When the Prophet called out to the Good Father of the Universe to remove his hand from the face of the Sun, the light gradually returned to the Earth. Response to the Prophet's performance was overwhelming and his fame spread rapidly. Harrison's condescension had backfired, to his embarrassment.

But the westward migration of European settlers was unstoppable. In the Battle of Tippecanoe in 1811, Harrison destroyed the Shawnee Prophet's religious center, killing many Indians and breaking the power of Tenskwatawa.*

---

*See especially Laurence A. Marschall, "A Tale of Two Eclipses," *Sky & Telescope* 57 (Feb. 1979): 116–118.

> meaning of the portent. They replied—"God is foreshadowing to the Greeks the destruction of their cities; for the sun foretells for them, and the moon for us." So Xerxes, thus instructed, proceeded on his way with great gladness of heart.[3]

To disaster! He reached and burned Athens, but his navy was destroyed by the Greeks and his forces had to withdraw. Twice more Xerxes invaded Greece, but each time his armies were crushed. After his last defeat, his nobles assassinated him.

Xerxes' first march against Greece actually occurred in 480 B.C., but the only major eclipse visible in the region near that date was the total eclipse of February 17, 478 B.C. Thus the story tells us less about observational astronomy in that era than about the power exercised by eclipses over the minds of men and the effectiveness of their use to heighten the drama of a story.

One final story illustrates the advance of the Greeks from superstitious dread of eclipses to an understanding of what causes them. On August 3, 430 B.C., Pericles and his fleet of 150 warships were about to sail for a raid upon their enemies.

> But at the very moment when the ships were fully manned and Pericles had gone onboard his own trireme, an eclipse of the sun took place, darkness descended and everyone was seized with panic, since they regarded this as a tremendous portent. When Pericles saw that his helmsman was frightened and quite at a loss what to do, he held up his cloak in front of the man's eyes and asked him whether he found this alarming or thought it a terrible omen. When he replied that he did not, Pericles asked, "What is the difference, then, between this and the eclipse, except that the eclipse has been caused by something bigger than my cloak?" This is the story, at any rate, which is told in the schools of philosophy.[4]

The eclipse was a large partial at Athens and annular about 600 miles (1,000 km) to the northeast. This eclipse had also been recorded by Thucydides, without the didactic story, but exhibiting an increased awareness of the cause of eclipses: "The same summer, at the beginning of a new lunar month, the only time by the way at which it appears possible, the sun was eclipsed after noon. After it had assumed the form of a crescent and some stars had come out, it returned to its natural shape."[5]

By 1654, Paris was a center of enlightenment, but on August 12, "at the mere announcement of a total eclipse, a multitude of the inhabitants of Paris hid themselves in deep cellars."[6]

No wonder then that the sight of a total eclipse on July 29, 1878, had a powerful effect on Native Americans near Fort Sill, Indian Territory (now Oklahoma). A non-Indian described it this way:

> It was the grandest sight I ever beheld, but it frightened the Indians badly. Some of them threw themselves upon their knees and invoked the Divine blessing; others flung themselves flat on the ground, face downward; others cried and yelled in frantic excitement and terror. Finally one old fellow stepped from the door of his lodge, pistol in hand, and, fixing his eyes on the darkened Sun, mumbled a few unintelligible words and raising his arm took direct aim at the luminary, fired off his pistol, and after throwing his arms about his head in a series of extraordinary gesticulations retreated to his own quarters. As it happened, that very instant was the conclusion of totality. The Indians beheld the glorious orb of day once more peep forth, and it was unanimously voted that the timely discharge of that pistol was the only thing that drove away the shadow and saved them from . . . the entire extinction of the Sun.[7]

The Animal Response

Noting their own primal response to the daytime darkening of the Sun, people through the ages have been fascinated by the reaction of animals to a total eclipse. Reports go back more than 750 years. In describing the eclipse of June 3, 1239, Ristoro d'Arezzo wrote: "[W]e saw the whole body of the Sun covered step by step . . . and it became night . . . and all the animals and birds were terrified; and the wild beasts could easily be caught . . . because they were bewildered."[8]

As a university student in Portugal, astronomer Christoph Clavius saw the total eclipse of August 21, 1560: "[S]tars appeared in the sky, and (miraculous to behold) the birds fell down from the sky to the ground in terror of such horrid darkness."[9]

In 1706 at Montpellier in southern France, observers reported that "bats flitted about as at the beginning of night. Fowls and pigeons ran precipitately to their roosts." In 1715, the French astronomer Jacques Eugène d'Allonville, Chevalier de Louville, traveled to London for the eclipse and observed that at totality "horses that were laboring or employed on the high roads lay down. They refused to advance."[10]

By 1842, some people were even conducting behavioral experiments on their pets. "An inhabitant of Perpignan [France] purposely kept his dog without food from the evening of the 7th of July. The next morning, at the instant when the total eclipse was going to take place, he threw a piece of bread to the poor animal, which had begun to devour it, when the sun's last rays disappeared. Instantly the dog let the bread fall; nor did he take it up again for two minutes, that is, until the total obscuration had ceased; and then he ate it with great avidity."[11]

William J. S. Lockyer, son of the pioneering solar spectroscopist, traveled to Tonga for an eclipse in 1911. The weather conditions were miserable and the insects numerous and very hungry. He and his colleagues caught only a brief view of the corona through thin clouds and the scientific results were very meager. The only members of his team with good results were those studying animal behavior. The horses did not seem to notice the darkening, but fowl ran home to roost and pigs lay down. Flowers closed. Most memorable of all were the insects, which had been completely silent until the moment of totality and then sang as if it were night. "The noise," recalled Lockyer, "was most impressive, and will remain in my memory as a marked feature of that occasion."[12]

"And it will come about in that day," declares the Lord God, "that I shall make the Sun go down at noon and make the Earth dark in broad daylight."
Amos 8:9

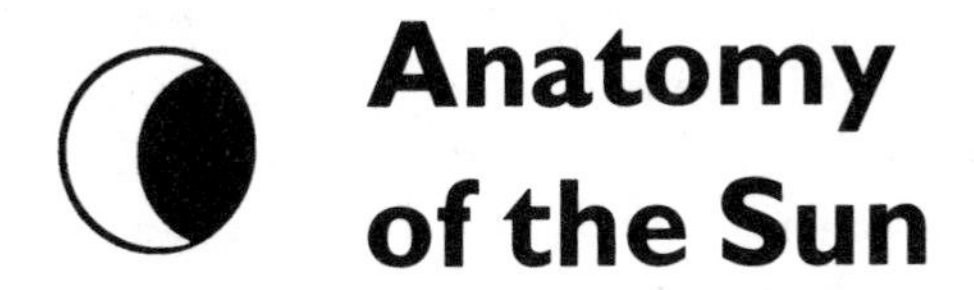

6 Anatomy of the Sun

At the Core

At this moment, deep in the core of the Sun, the nucleus of a hydrogen atom—a proton—is colliding and fusing with another hydrogen nucleus, and the collisions and fusions proceed until four hydrogen nuclei have become the nucleus of one helium atom. In this nuclear reaction, a tiny amount of mass has been destroyed: not lost, but converted into energy. It is this reaction that powers the Sun and all stars, creates their light and heat, for more than 90 percent of their active lives.

In the Sun, tens of trillions of these reactions take place every *second.* Every second 700 million tons of hydrogen become 695 million tons of helium, and 5 million tons of mass become energy, in accordance with Einstein's famous equation $E = mc^2$. A little bit of mass yields a vast amount of energy.[1]

Even though the Sun is actually losing mass at the rate of 5 million tons every second, this weight-reduction plan is far from a crash diet. At 5 million tons a second, it would take the Sun 117 billion years to consume itself entirely, if it could. But it can't. The heat to run this nuclear fusion comes from the gravitational force of all the mass of the Sun pressing inward on the core, and it is only within this central 25 percent of the Sun's diameter that the temperatures—up to 27 million degrees Fahrenheit (15 million degrees Celsius)—are hot enough to generate and sustain this reaction.[2]

Over billions of years, the hydrogen at the core is converted into helium until the core is too clogged with helium for hydrogen fusion to continue. It is then that stars begin to die. Our Sun has been shining for 4.6 billion years and has enough hydrogen at its core to continue shining much as it does now for about another 5 billion.

The fiercely hot reaction at the heart of our Sun is concealed from our view by 432,500 miles (696,000 km) of opaque gases. And a good thing too. The principal radiation generated at the Sun's core is not visible light but gamma rays. There would be no life on Earth if the Sun radiated large quantities of this high-energy radiation.

Fortunately, it does not. High-energy photons radiate outward from the Sun's core but are absorbed in the crush of other atomic particles, which in turn reradiate that energy. But the energy emitted is not the same. With each absorption and emission, some of the pho-

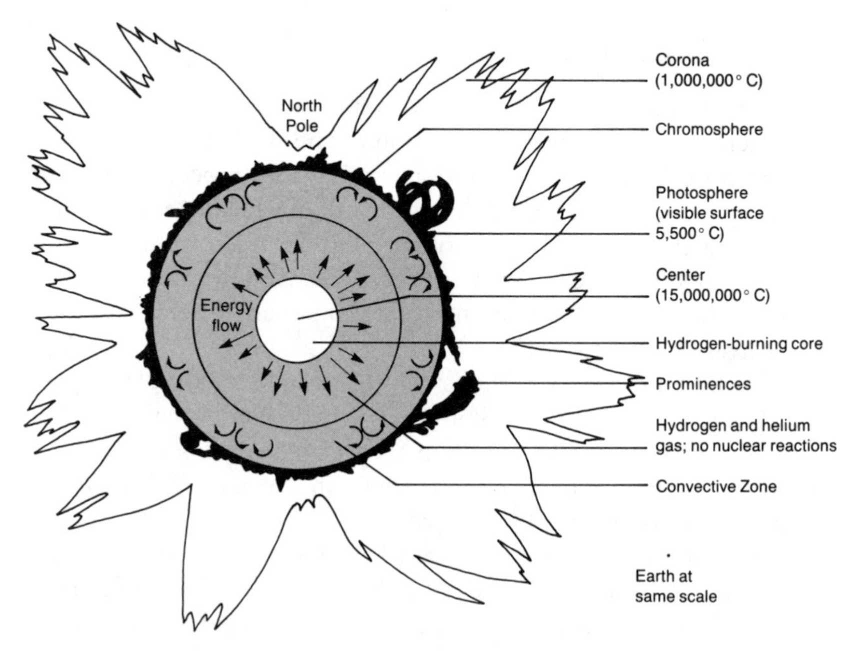

Cross section of the Sun and its atmosphere. *Drawn by Josie Herr after William K. Hartmann,* Cosmic Voyage Through Time and Space, *© 1990 Wadsworth Publishing, Inc., by permission of the publisher.*

ton's original energy is lost. A photon that started as a highly energetic gamma ray, if tracked from absorption to absorption, would gradually become an x-ray, then an ultraviolet ray, and then visible light as it bounced around randomly inside the Sun. At about 130,000 miles (200,000 km) below the Sun's surface, the temperature and density have fallen enough so that energy is conveyed upward less by radiation than by convection—the rise of gases heated by energy from below.

If energy created at the Sun's core could leave the Sun directly traveling at the speed of light, it would emerge from the Sun's surface in 2⅓ seconds. But because of the countless absorptions and reemissions in random directions, a typical photon requires 10 million years to reach the Sun's surface. There, at last, it is free to move directly away from the Sun at the speed of light. At that pace, it travels the distance from the Sun to the Earth in 8⅓ minutes.

Layers Above

All the Sun's internal layers of radiation and convection are hidden from our eyes. What we see of the Sun, when we or the atmosphere provide adequate filters, is an apparent disk, glaring white-hot at a temperature of about 10,000 °F (5,500 °C). This disk is an optical illusion, since the Sun is gaseous throughout. We are really seeing the layer of the Sun in which the density and ionization of atoms is so great that the gas becomes opaque. This region that provides the Sun with the appearance of a surface is called the *photosphere* ("light sphere"). It is only about 200 miles (300 km) deep. It is there that we notice sunspots, areas of magnetic disturbance on the Sun the size of Earth or Jupiter or even larger, appearing and disappearing and riding along in the photosphere with the Sun's rotation.

Above the photosphere is the *chromosphere* ("color sphere"), aptly named for its vibrant reddish color. Seen with a telescope at the rim of the Sun, it looks like a fire-ocean.[3] As its color suggests, its lower levels are cooler than the white-hot photosphere, with temperatures of about 7,200 °F (4,000 °C). The chromosphere is a thin atmospheric layer, only about 1,600 miles (2,500 km) thick, although there are no sharp boundaries above or below.

It is in the chromosphere that prominences and flares are rooted and stretch upward into the corona. They are additional transient

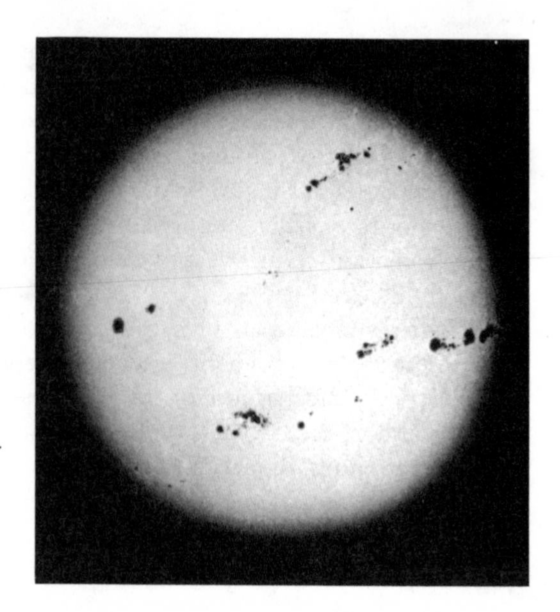

Photosphere of the Sun at sunspot maximum (December 1957). *Carnegie Observatories*

features of the Sun that vary with the rhythm of the sunspots below them. *Prominences* are condensed clouds of solar gas, slightly cooler than their surroundings, bent and twisted by local magnetic fields. Most often these clouds rain material back toward the surface, but occasionally they erupt outward. *Flares* are much stronger eruptions, also triggered by the magnetic activity of the Sun, that launch great torrents of mass and energy from the Sun at millions of miles an hour.

Above the chromosphere lies the *corona* ("crown"), the outer atmosphere of the Sun. It is white in color and, peculiarly, it is extremely hot, with variable temperatures often exceeding 2 million °F (1 million °C). This temperature can be misleading. The Sun's magnetic fields cause the electrically charged atoms of the corona to move at great speeds (high temperature), but the density of these ions is so low that the corona has relatively little heat (energy in a given volume). The corona is so rarefied that if you had a box there 100 miles (160 km) on each side (1 million cubic miles; 4.1 cubic kilometers), you would entrap less than a pound (0.4 kilogram) of matter. The corona is a good vacuum by Earth laboratory standards.

To the eye, the white-flame brushes of the corona can extend outward from the Sun 2 million miles (3 million km) or more until they are so tenuous that they are no longer visible. But the corona is still measurable to the Earth and beyond. The Earth orbits the Sun within the Sun's tenuous outer atmosphere.

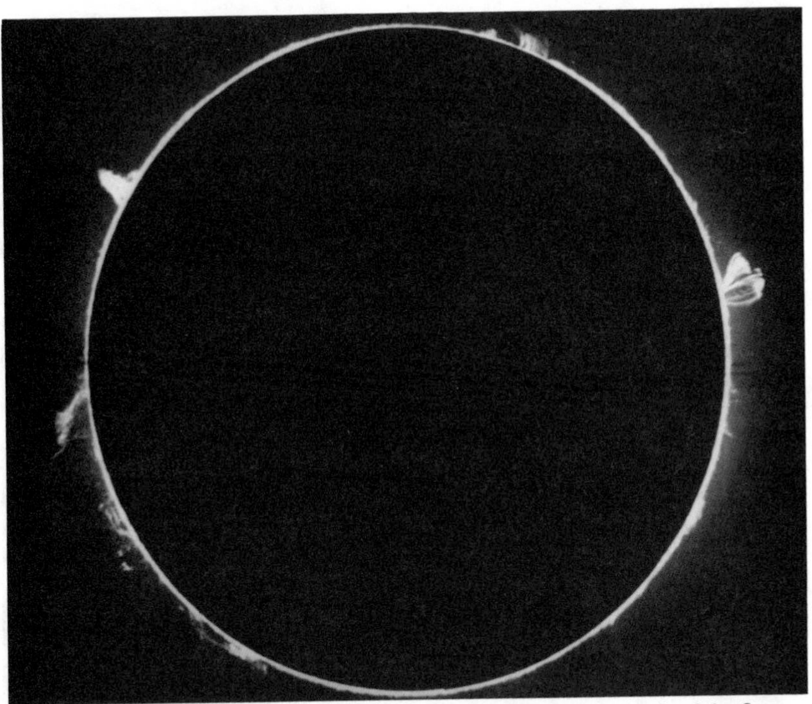

Prominences of different sizes and shapes silhouetted against the limb of the Sun. *Carnegie Observatories*

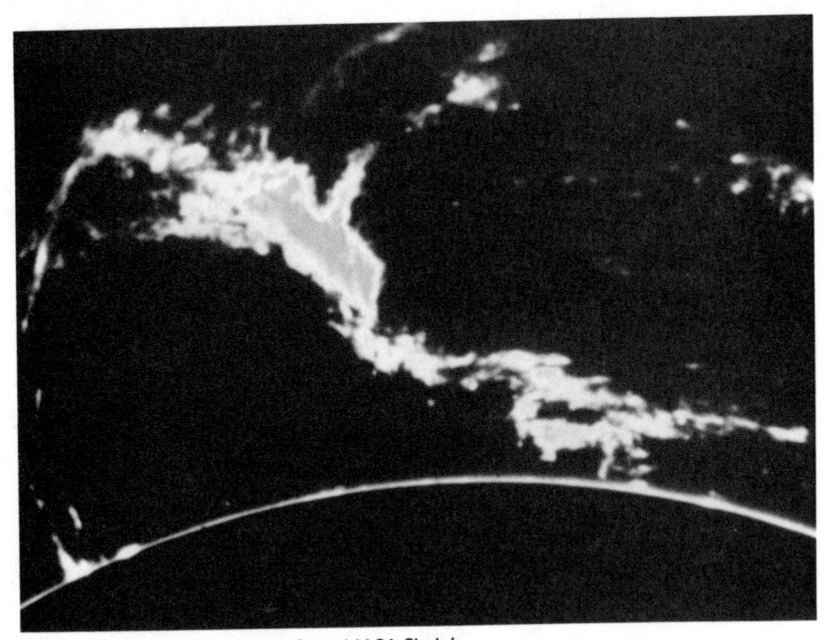

Eruptive prominence on the Sun. *NASA Skylab*

Corona (June 8, 1918). *Lick Observatory*

We do not normally see the chromosphere or the corona. They are concealed from our view by the overwhelming glare of the photosphere, half a million times brighter than the corona. To study the Sun, early scientists had only its surface to rely on—just the photosphere and sunspots. Progress was slow.

We do not see the chromosphere or the corona unless something blocks the glare of the photosphere, allowing the faint atmosphere of the Sun to be revealed. In the nineteenth century, astronomers discovered that the Moon was their scientific collaborator, obscuring the Sun's surface from time to time so that they might see a part of the Sun that had never been studied before. In less than a century, total solar eclipses, and the scientific instruments and theories they helped to stimulate, revealed the composition of the Sun, the structure of the Sun's interior, and the wonder of how the Sun shines.

. . . I did not expect, from any of the accounts of preceding eclipses that I had read, to witness so magnificent an exhibition as that which took place.
Francis Baily (1842)

7

Lessons from Eclipses

An Unlikely Beginning

Francis Baily, the man who might be said to have founded the field of solar physics, received only an elementary education, was not trained in science, and did not get around to astronomy until the age of thirty-seven. Like his father, a banker, he entered the commercial world as an apprentice when he was fourteen. But adventure called. When his seven years of apprenticeship expired, he sailed for the New World and spent the next two years, 1796–1797, exploring unsettled parts of North America, narrowly escaping from a shipwreck, flatboating down the Ohio and Mississippi rivers from Pittsburgh to New Orleans, and then hiking nearly 2,000 miles back to New York through territory inhabited mostly by Indians. He liked the United States so well that he planned to marry and become a citizen, but he finally abandoned those plans and returned home in 1798.

Back in England, he began efforts to mount an expedition to explore the Niger River in Africa. He could not raise enough money, however, so he became a stockbroker. To dedication and enthusiasm he quickly added a reputation for intelligence and integrity, and he made a fortune. He exposed stock-exchange fraud and helped clean it up. He published a succession of explanations of life insurance methods and comparisons of insurance companies, which became wildly popular. He also published a chart of world history that was equally popular, confirming the nickname given to him in his apprentice days: The Philosopher of Newbury (his birthplace).

His first astronomical paper (1811) tried to identify the solar eclipse allegedly predicted by Thales. In 1818, he called attention to an annular eclipse of the Sun coming in 1820, and he observed it from southeastern England. That same year, he became one of the founders of the Astronomical Society of London, later the Royal Astronomical Society. In 1825, he retired from the stock market to devote all his time to his new profession. He was 51 years old. His revisions of a series of old star catalogs were considered so valuable that the Royal Astronomical Society twice awarded him its Gold Medal and four times elected him president. Although he was not renowned as an observer, he had an abiding fascination with eclipses, a good eye for detail, and the ability to express what he saw.

Thus it was in 1836 that a few words from Francis Baily sparked the immediate, intense, and unending study of the physical properties of the Sun that had been generally ignored or discounted until then. He traveled to an annular eclipse of the Sun in southern Scotland and watched on May 15, 1836, as mountains at the Moon's limb occulted the face of the Sun but allowed sunlight to pour through the valleys between them so that the ring of sunlight around the rim of the Moon was broken up into "a row of lucid points, like a string of bright beads."[1] With those words, Baily generated fervor for solar physics and founded the industry of eclipse chasing.

The Surprise of Totality

At the next accessible eclipse, July 8, 1842, a high percentage of the astronomers of Europe migrated to southern France and northern Italy to see "Baily's Beads." Baily, sixty-eight years old, went too. This was not an annular eclipse, as Baily had seen twice before. It was total. No European astronomer then alive had ever seen a total eclipse.

Baily set up his telescope at an open window in a building at the university in Pavia, Italy: "[A]ll I wanted was to be left *alone* during the whole time of the eclipse, being fully persuaded that nothing is so injurious to the making of accurate observations as the intrusion of unnecessary company."[2]

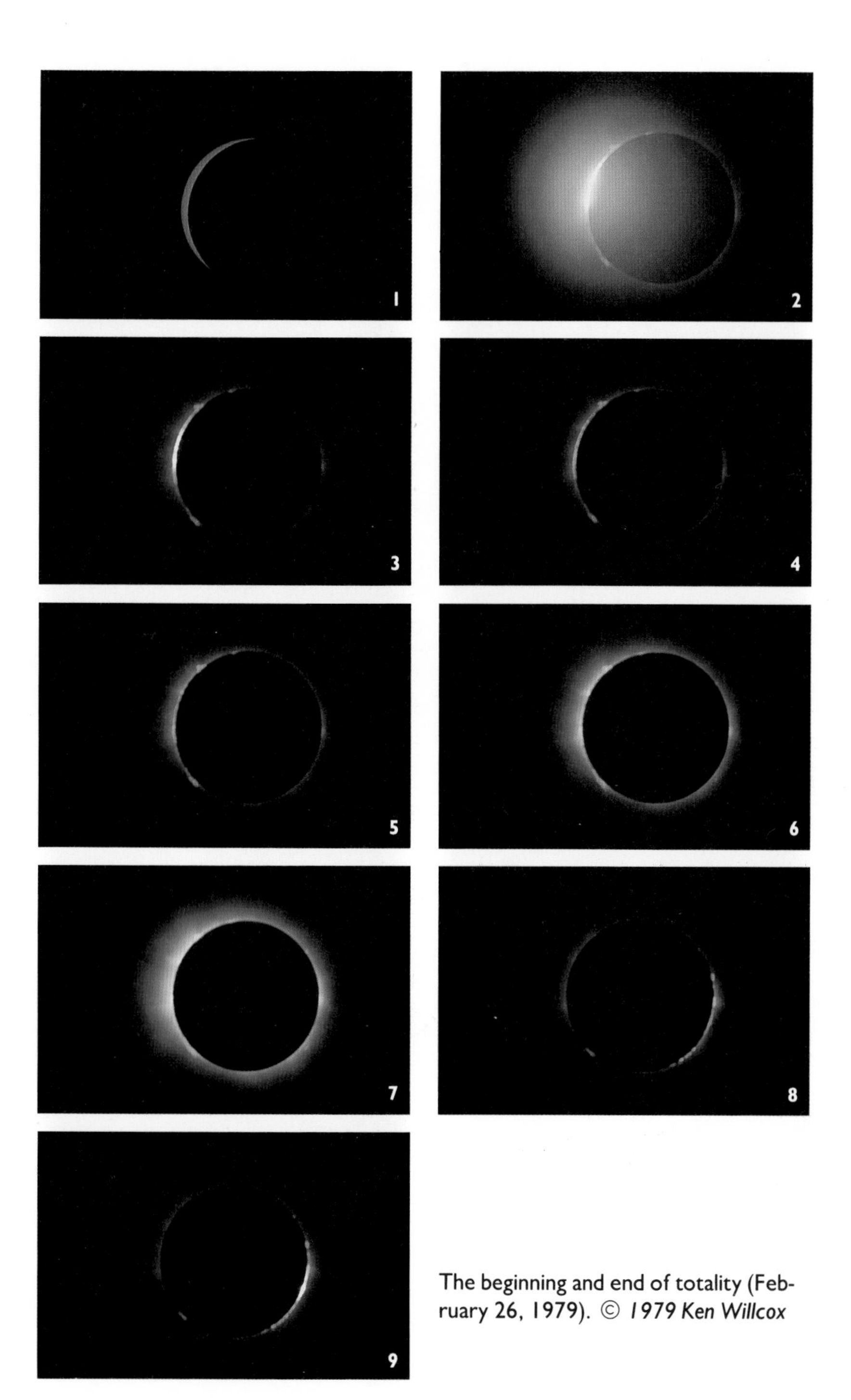

The beginning and end of totality (February 26, 1979). © *1979 Ken Willcox*

Drawing by Lilian Martin-Leake from a telescopic view of the chromosphere and corona (May 28, 1900), showing red spicules in the chromosphere that George Airy had thought were mountains.

Opposite page: Diamond Ring Effect with two rubies (November 12, 1966). *NASA*

Annular eclipse showing chromosphere, Baily's Beads, and a remarkable prominence (May 30, 1984). © *1984 Dennis di Cicco*

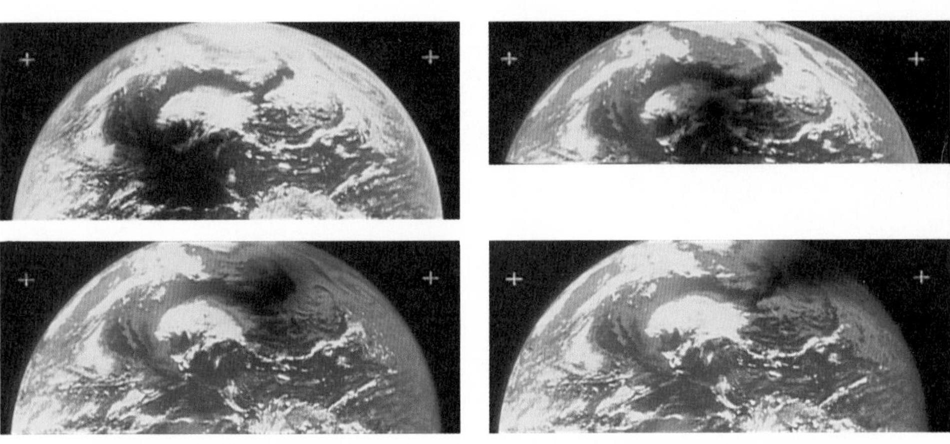

From top left: The dark shadow of the Moon moves northeastward over the southeastern United States and the North Atlantic during a total eclipse seen by NASA's ATS-3 weather satellite (March 7, 1970). *NASA*

The Moon's dark shadow on the horizon, as seen from an aircraft above the clouds (July 30–31, 1981). © *1981 Stephen J. Edberg*

Twilight colors on the horizon beyond the Moon's shadow during an eclipse in India (February 16, 1980). © *1980 Jay M. Pasachoff*

An x-ray image of the Sun's disk superimposed on a white light photograph of the corona taken during an eclipse in the Philippines (March 17–18, 1988). The x-ray image was taken about six hours earlier using a NASA sounding rocket. Where the edge of the Sun's disk is brighter (more active), the corona extends farther into space. *NCAR and Laboratory for Atmospheric and Space Physics/University of Colorado*

The Earth eclipses the Sun as seen by Apollo 12 astronauts en route to the Moon in 1969. From their position the apparent disk of the Earth was much larger than the apparent disk of the Sun. *NASA*

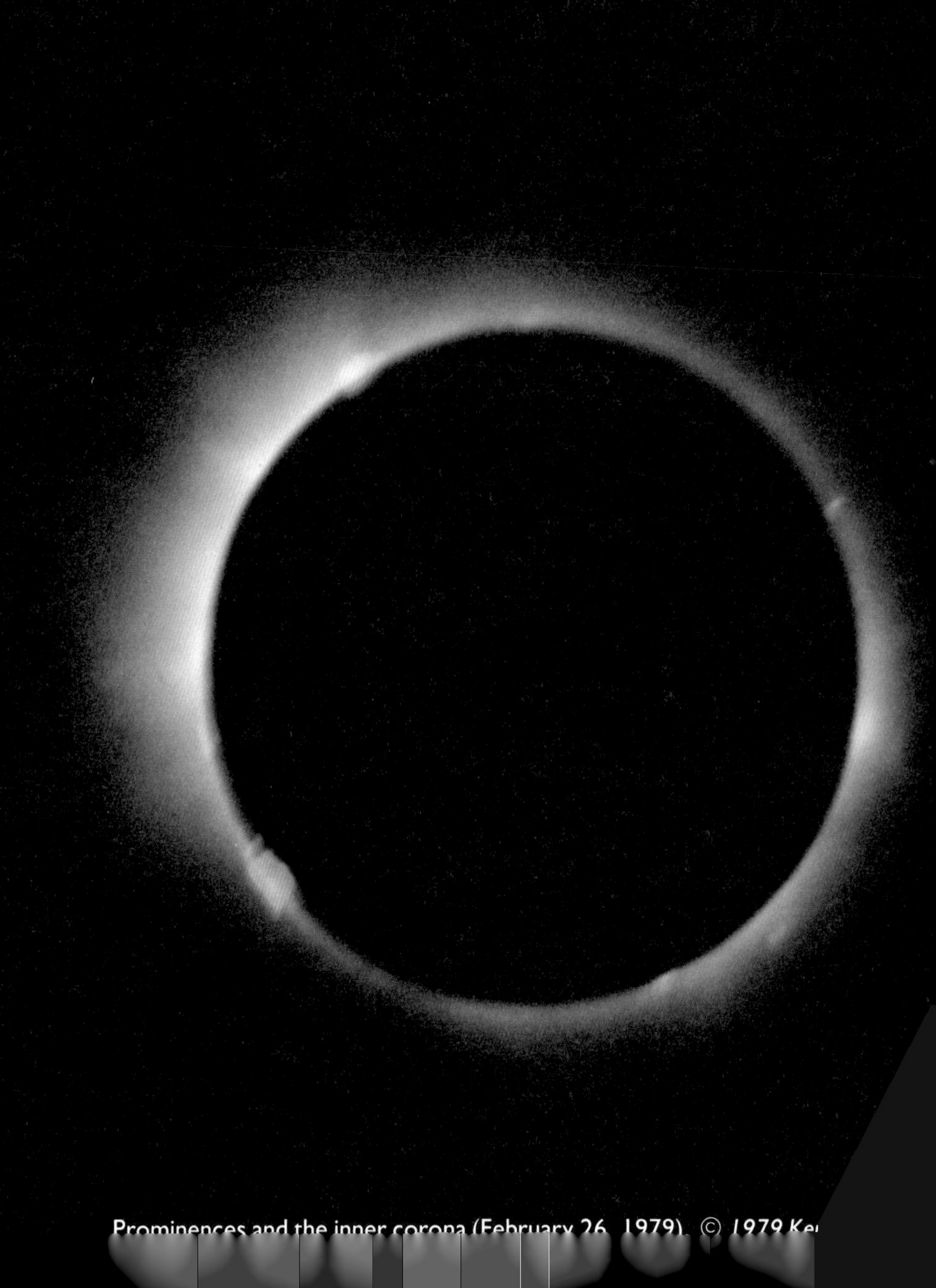

Prominences and the inner corona (February 26, 1979). © 1979 Ke

By 1860, daguerreotypes were obsolete, superseded by the faster wet plate (collodion) photographic process. Before photography, astronomers could only describe or sketch what they saw. The early photographic emulsions were not as sensitive to detail as desired, but they were more objective than the human eye.

On July 18, 1860, a total eclipse was visible from Europe and found astronomers waiting with improved cameras. Prominences remained a matter of high priority and they encountered the resourcefulness of British astronomer Warren De La Rue and Italian astronomer and Jesuit priest Angelo Secchi.

De La Rue's well-to-do family provided him with a solid education, after which he entered his father's printing business. There he demonstrated a great affinity for machinery. He could make any instrument or device run better. He was also a fine draftsman, and it was this talent that lured him into astronomy. He could produce drawings of the planets, Moon, and Sun that were better than those of astronomers. But no sooner had he produced his first excellent drawings than he found out about a new invention called photography. As he never saw a mechanical device that did not fascinate him, he was soon making improvements in cameras, inventing specialized cameras for solar photography, and photographing the Moon and Sun stereographically so that lunar features appeared in relief and sunspots revealed themselves as depressions, not mountains, in the photosphere.[7]

Observing the 1860 eclipse 250 miles (400 km) away from De La Rue was Angelo Secchi. He too had a remarkable aptitude for inventing and coaxing instruments. Secchi was from a poor family and received his education through the Catholic Church in the demanding Jesuit tradition. From the beginning, he showed brilliance in mathematics and astronomy. When the Jesuits were expelled from Italy by a liberal, anticlerical government in 1848, Secchi spent a year in the United States as assistant to the director of the Georgetown University observatory. When the ban against Jesuits was lifted in 1849, he returned to Italy to become director of the Pontifical (now Vatican) Observatory of the Collegio Romano (or Gregorian University). He transformed it into a modern, well-equipped center for research in the new field of astrophysics. Secchi was one of the pioneers in applying spectroscopy to astronomy. He surveyed more than 4,000 stars and realized that all stellar spectra could be grouped into a handful of classifications.

Angelo Secchi

De La Rue and Secchi recorded the 1860 eclipse with improved cameras using the wet-plate process, which greatly reduced exposure time and thereby increased the clarity with which objects in motion could be seen. They captured the prominences and compared them. The prominences looked the same from the photographers' widely separated sites, so Secchi and De La Rue could conclude that they were indeed part of the Sun. If they had been features on the Moon, so much closer than the Sun, the difference in viewing angles (parallax) from Secchi's and De La Rue's separate sites would have given them a different appearance.

The Debut of Spectroscopy

On August 18, 1868, a great eclipse touched down near the Red Sea and swept across India and Malaysia. Once again, the international scientific community had assembled.

From the United Kingdom to the path of the eclipse had come, among others, James Francis Tennant and John Herschel (son of John

F. W. Herschel, grandson of William Herschel, both renowned astronomers). Norman Pogson, born in England, represented India and the observatory he directed there. The French delegation included Georges Rayet and Jules Janssen. Each of them carried a new weapon just added to the scientific arsenal for prying secrets from the Sun during an eclipse. That tool was a spectroscope. It would prove as indispensable to eclipse studies of the structure of the Sun as it was rapidly demonstrating itself to be in other realms of astronomy and in all other physical and biological sciences. By passing the light of the corona or the prominences through a prism, it could be broken down into a spectrum of lines and colors. From this spectrum, a scientist could identify the chemical elements present and even the temperature and density of its source.

As the eclipse sped along its course, spectroscopes pointed upward toward the prominences. They showed that the prominences emitted bright lines, and most of these were quickly identified with hydrogen. More and more it seemed that the Sun must be composed primarily of gas and that hydrogen must be a major constituent.[8] The spectroscopists, properly pleased with their results, packed their equipment and headed home. All but one. His name was Jules Janssen and the brightness of the prominences and the strength of their spectral lines had given him an idea. He wanted to look for them again when the eclipse was over, when the Moon was not blocking the intense glare of the Sun from view. Might it be possible for him to see the prominences and their spectrum in broad daylight?

The weather was cloudy the rest of that day. He would have to wait until tomorrow.

The Extra Step

Pierre Jules César Janssen was twelve years old when Baily called attention to the beads visible during the annular eclipse of 1836. A childhood accident had left Janssen lame and he never attended elementary or high school. His family was cultured, but his father was a struggling musician, so Jules had to go to work at a young age.

While employed at a bank, he earned his college degree in 1849. He then went on to gain a certificate as a science teacher and served as a substitute teacher at a high school. In 1857, he traveled to Peru as part of a government team to determine the position of the magnetic

Jules Janssen. *Mary Lea Shane Archives of the Lick Observatory*

Jules Janssen
(1824–1907)

Two years after his breakthrough at the 1868 eclipse, Jules Janssen again planned to apply spectral analysis to a solar eclipse, this one on December 22, 1870, in Algeria. When it was time for departure, however, the Franco-Prussian War was in progress and Paris was under siege. Colleagues in Britain had obtained from the Prussian prime minister safe passage for Janssen from Paris, but Janssen would not accept favors from his country's enemies. He had a different plan: "France should not abdicate and renounce taking part in the observation of this important phenomenon. . . . an observer would be able, at an opportune moment, to head toward Algeria by the aerial route . . ."*

Although he had never been in a balloon before, on December 2, with a sailor as an assistant and himself as pilot, he ascended from Paris and headed west. Despite violent winds, he landed safely near the Atlantic coast. He reached Algeria in time—only to have the eclipse clouded out. From that experience, however, Janssen designed an aeronautical compass and ground speed indicator and prophesied methods of air travel that would "take continents, seas, and oceans in their stride." In 1898 and subsequently, he used balloons to study meteor showers from above the

equator. There he became severely ill with dysentery and was sent home. At age thirty-three, he seemed destined for a quiet life in teaching, if he could get a job. He became the tutor for a wealthy family in central France. At their steel mills, he noticed that the eye could watch molten metal without fatigue or injury, while the skin had to be protected from the heat. He wrote a careful study of how the eye protects itself against heat radiation, which earned him his doctorate in 1860.

The glow of molten metal had led him to spectroscopic analysis, which he then applied to the Sun and, in 1859, identified several Earth elements present in the Sun. He moved to Paris in 1862 to dedicate himself to solar physics and scientific instrument-making. His work had already made him a leader in solar spectroscopy. He used the changing spectrum of the Sun in its daily journey from horizon to horizon to separate spectral lines caused by the Earth's atmosphere from those originating in the Sun and to demonstrate the composition and density of the Earth's atmosphere through which the sun-

---

clouds, pioneering high-altitude astronomy and foreseeing the advantages of space observations.

Janssen was also a leader in astrophotography. "[T]he photographic plate is the retina of the scientist," he wrote. It was for spectroscopy, however, that Janssen was most renowned. His methods opened up the Sun's atmosphere to continuous study. The French government tried to find him an observatory position, but the director of the Paris Observatory did not want him. Janssen was allowed to pick a site near Paris for a new observatory dedicated to astrophysics. He chose Meudon and directed the observatory from its founding in 1876 until his death in 1907.

"There are very few difficulties that cannot be surmounted by a firm will and a sufficiently thorough preparation," he wrote. But he was too modest in his self-assessment. To everything he investigated, he brought imagination and insight. Jules Janssen was one of the most creative scientists of any era.

---

*Quotations are from the sketch of Janssen by Jacques R. Lévy in the *Dictionary of Scientific Biography*.

---

light passed. Janssen then applied spectroscopy to the other planets and, in 1867, discovered water in the atmosphere of Mars.

He traveled to Guntur, India for the total eclipse of August 18, 1868, to use his spectroscope on solar prominences. The contemporary English spectroscopist Norman Lockyer spoke ringingly of that pivotal moment: "Janssen—a spectroscopist second to none— . . . was so struck with the brightness of the prominences rendered visible by the eclipse that, as the sun lit up the scene, and the prominences disappeared, he exclaimed, '*Je reverrai ces lignes là!*' [I will see those lines again!]"[9] The next morning he succeeded. He had found a way to study the atmosphere of the Sun without waiting for a total eclipse, traveling halfway around the world to see it, and hoping for good weather at the immutable moment.

For two weeks Janssen continued to map solar prominences by this technique and continued to perfect it on his circuitous way home, with a stop in the Himalayas to observe at high altitude. He proved that prominences change considerably from one day to the next. A standard spectroscope breaks down the light of a glowing object into the characteristic colors of its spectrum. Janssen modified the spectroscope by blocking unwanted colors so that the observer could view an object in the light of one spectral line at a time. He had invented the spectrohelioscope. The Sun could now be analyzed in detail on a daily basis.

A month after the eclipse, on his way home to France, Janssen wrote up his findings and sent them to the Academy of Sciences in Paris. His paper arrived a few minutes after one from England that reported precisely the same discovery.

Coincidence

Joseph Norman Lockyer came from a well-to-do family with scientific interests. He received a classical education, traveled in Europe, and then entered civil service. So wide were his interests that he wrote on everything from the construction dates and astronomical purposes of Egyptian pyramids and temples to Tennyson to the rules of golf. "The more one has to do, the more one does," was his motto.[10]

When Gustav Kirchhoff and Robert Bunsen showed in 1859 how spectroscopy could be used to determine the chemical composi-

J. Norman Lockyer in 1895

tion of objects in space, Lockyer saw the discovery as a key to what had seemed the locked door of the universe. Had not French philosopher Auguste Comte confidently asserted only 24 years earlier that never, by any means, would we be able to study the chemical composition of celestial bodies, and every notion of the true mean temperature of the stars would always be concealed from us.[11] It was clear that Comte was wrong. Lockyer bought a spectroscope, attached it to his 6¼-inch (16-cm) refracting telescope and began his observations.

Although he had never seen a solar eclipse, it occurred to him that, since prominences were probably clouds of hot gas, he should be able to use a spectroscope to analyze prominences without waiting for an eclipse. This idea struck Lockyer two years before the 1868 eclipse that inspired Janssen to the same realization. Lockyer tried the experiment in 1867 but found his spectroscope inadequate to the task. So he ordered a new spectroscope to his specifications. Because of construction delays, however, it did not arrive until October 16, 1868, two months after the eclipse that Janssen saw in India. Lockyer rapidly and excitedly calibrated his new instrument and on October 20, 1868, he trained it on the rim of the Sun and recorded bright lines

Medallion created by the Academy of Sciences in Paris to honor Janssen and Lockyer. The front of the medal shows the heads of the two scientists; the reverse shows the Sun god Apollo pointing to prominences on the Sun.

typical of hot gases under very little pressure. He wrote up his findings and sent them to the Academy of Sciences in Paris for presentation by his friend Warren De La Rue.

Just minutes before De La Rue was to speak, Janssen's letter arrived and both papers were read at the same session of the Academy to great acclaim for both scientists. A special medal was struck to honor them. It showed the heads of Janssen and Lockyer side by side.[12]

J. Norman Lockyer
(1836–1920)

In 1869, the year after he independently showed how the atmosphere of the Sun could be analyzed without the benefit of an eclipse and discovered the element helium, Lockyer founded the scientific journal *Nature*. He edited it for 50 years, keeping it alive through many crises.

While the French government was establishing a special observatory for Janssen, the British government likewise recognized the importance of Lockyer's contribution and set about creating a solar physics observatory for him. For its opening in 1875, Lockyer collected old and modern instruments and placed them on display. The display became permanent and grew. Lockyer had founded London's world-famous Science Museum.

Lockyer was not shy in interpreting his findings to form startling theories. Often he was wrong, but always he provided useful data and frequently there was a nucleus of truth in his grand speculations. He thought that all atoms shared certain spectral lines and were therefore

A New Element

Lockyer continued to examine the spectrum of the gases at the rim of the Sun. He recognized that the lower atmosphere of the Sun, what Airy had called the sierra, was decidedly reddish in color, so he named it the *chromosphere*, and it has been known by that name ever since.

Lockyer was not done yet. In examining the spectrum of the prominences, he noticed a yellow line that he could not identify. It did not seem to belong to any element known on Earth. So he announced the existence of a new element and proposed the name *helium* for it, because it had been found in the Sun—*helios* in Greek. Most scientists rejected the idea of a new element, suggesting that this line was produced by a known element under unusual physical conditions. But Lockyer clung tenaciously to his interpretation. Finally, in 1895, William Ramsay found trapped in radioactive rocks on Earth an unknown gas that exhibited the mysterious spectral line that Lockyer had discovered on the Sun. Helium was an element. Lockyer had been right. In 1869, just after he discovered helium, Lockyer had urged: "[L]et us . . . go on quietly deciphering one by one the letters of this strange hieroglyphic language which the spectroscope has revealed to us—a language written in fire on that grand orb which to us earth-dwellers is the fountain of light and heat, and even of life itself."[13]

---

made of smaller common constituents. He was wrong about the spectra, but ultimately right about the composition of atoms.

He offered dates for the construction of ancient Egyptian temples based on their alignments with the rising and setting positions of the Sun and certain stars. His dates were wrong, but he was right that many of the temples had astronomical orientations, and his work helped to establish the field of archeoastronomy.

When he died, a colleague wrote of him: "Lockyer's mind had the restless character of those to whom every difficulty is a fresh inspiration. His enthusiasm never failed him, despite repeated disappointments and opposition."*

*Alfred Fowler, "Sir Norman Lockyer, K.C.B., 1836–1920," *Proceedings of the Royal Society of London,* ser. A, 104 (1923): i–xiv.

The total eclipse of 1868 had raised the strong possibility of the existence of a new element—and, for the first time, one discovered not on Earth but in the heavens. One year later, on August 7, 1869, the United States lay on the path of a total eclipse. Two American astronomers, Charles A. Young and William Harkness, working separately, observed the event with spectroscopes. Each noticed a green line in the spectrum of the corona that defied identification with known elements. This suspected new element was called *coronium*.

In 1895, Lockyer's helium was identified in rocks on Earth, but coronium remained a spectral presence seen only on the Sun during total eclipses. As time passed, more elements were discovered on Earth until the periodic table of stable chemical elements was nearly complete. There was no room left for coronium to be an element. What could it be?

Might it be an already known element under such unusual conditions that it emitted a spectrum never before seen in a laboratory? Walter Grotrian of Germany in 1939 pointed the way and Bengt Edlén of Sweden in 1941 identified the green line of coronium as the element iron with 13 electrons missing—a "gravely mutilated state."[14] To ionize iron so greatly, the temperature of the corona had to be about 2 million °F (1 million °C) and its density had to be less than a laboratory vacuum. Because the conditions necessary for the production of such lines cannot be achieved in a laboratory, they are known as forbidden lines.

The Reversing Layer

Astronomy was a family tradition for Charles Augustus Young. His maternal grandfather and his father had been professors of astronomy at Dartmouth. Charles entered Dartmouth at age fourteen and four years later graduated first in his class. He immediately began teaching—classics!—at an elite prep school, and commenced studies at a seminary to become a missionary. In 1856 he changed his plans and became professor of astronomy at Western Reserve College, with a break in his duties to serve in the Civil War. He returned to Dartmouth in 1866, holding the same chair as his father and grandfather before him. There he pioneered in spectroscopy, especially applied to the Sun.

At the eclipse of December 22, 1870, which he observed at Jerez, Spain, Young noticed that the dark lines in the Sun's spectrum

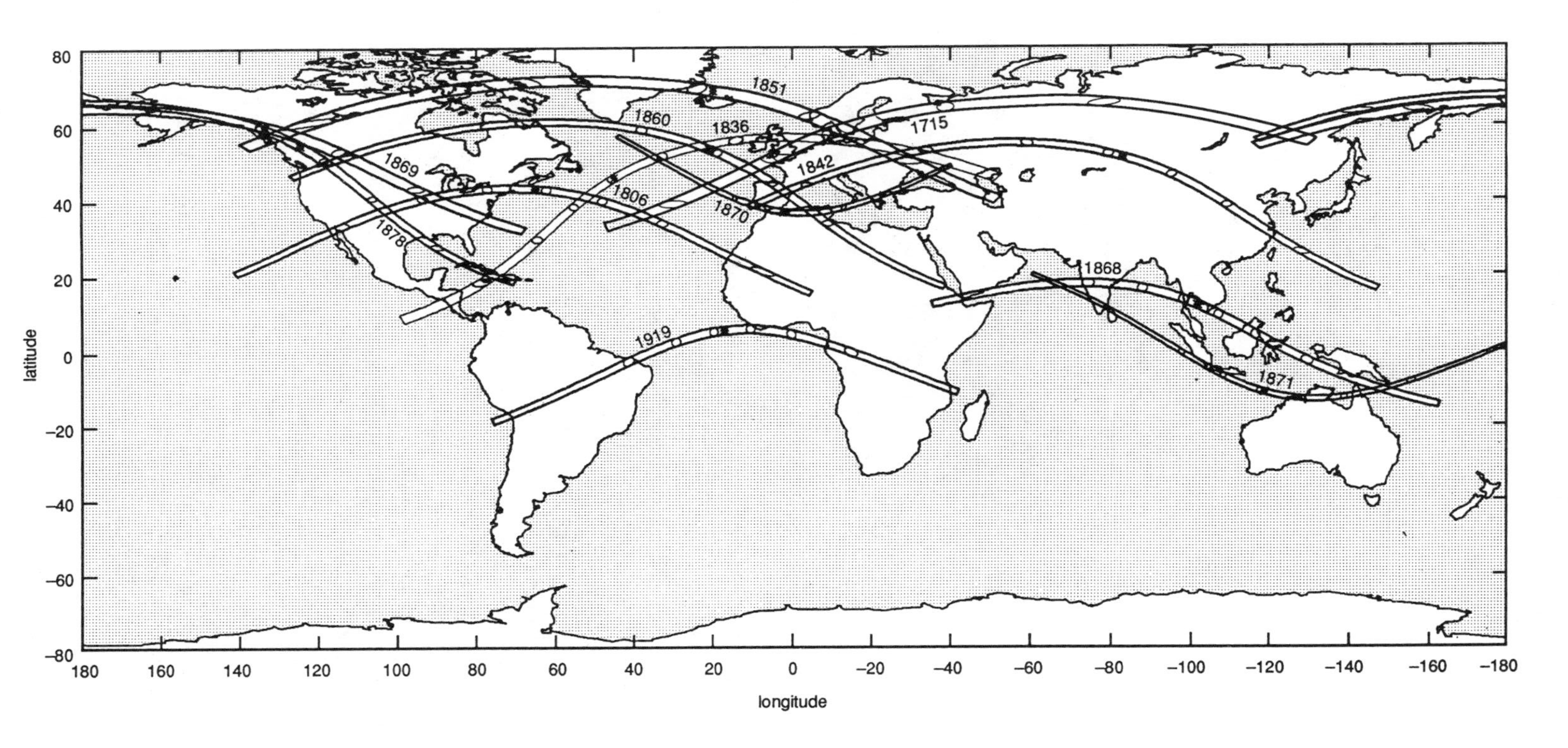

Paths of totality for the solar eclipses of 1715, 1806, 1836, 1842, 1851, 1860, 1868, 1869, 1870, 1871, 1878, and 1919. *Map and eclipse calculations by Fred Espenak, NASA Goddard Space Flight Center*

Charles A. Young. *Mary Lea Shane Archives of the Lick Observatory*

Lessons from Eclipses at Other Planets
by Carl Littmann

As the moons in the solar system revolve around their planets, they too create and undergo periodic eclipses. These events allowed the Danish astronomer Ole Römer in 1676 to prove, contrary to prevailing opinion, that light travels at a finite speed. He even succeeded in making the first good estimate of the speed of light. Such luminaries as Aristotle, Kepler, and Descartes had been certain that the speed of light was infinite. Gian Domenico Cassini, director of the Paris Observatory where Römer made his discovery, refused to believe the results, and Römer's triumph was not fully appreciated for half a century.

In observations of Io, innermost of Jupiter's four large moons, Römer noticed discrepancies between the observed times of its disappearance into the shadow of the planet and the calculated times for these events. He correctly explained that these discrepancies were due to the travel time of light between Jupiter and the Earth. When the Earth is approaching Jupiter, the interval between satellite eclipses is shorter because the distance light must travel is decreasing. When the Earth is moving farther from Jupiter, the interval between eclipses lengthens because the distance light must travel is increasing. Römer determined that light from Io took about 22 minutes longer to reach the Earth when the Earth was farthest from Jupiter than when it was closest. Thus light required about 22 minutes to cross the orbit of Earth. The diameter of the Earth's orbit was not well known at the time. Modern measurements show that light actually requires about 16⅔ minutes to make this journey.

When Römer returned to Denmark, the king gave him a succession of appointments, including master of the mint, chief judge of Copenhagen, chief tax assessor (everyone said he was fair!), mayor and police chief of Copenhagen, senator, and head of the state council of the realm—all this and more while he served as director of the Copenhagen observatory and astronomer royal of Denmark. He discharged all his duties with distinction.

Solar eclipse on Jupiter. Black dot at the left is the shadow of Ganymede, Jupiter's largest moon. To the right is Io, Jupiter's innermost large satellite. *NASA/JPL*

Solar eclipse on Saturn caused by its ring system. At the top are moons Tethys and Dione. Tethys' shadow on Saturn can be seen at the lower left, just above the shadow of the rings. *NASA/JPL*

Saturn's shadow eclipses its rings. *NASA/JPL*

become bright lines for a few seconds at the beginning and end of totality. He had discovered the *reversing layer,* the lowest 600 miles (1,000 km) of the chromosphere, which is cooler than the photosphere and thus absorbs radiation of specific wavelengths, producing the ordinary dark-line spectrum of the Sun. However, when the Moon blocks the photosphere from view and the reversing layer of the chromosphere can be seen momentarily before it too is eclipsed, the bright-line spectrum of its glowing gases flashes briefly into view. Here at last was the layer responsible for the dark-line spectrum of the Sun seen on ordinary days. A long-missing piece in the puzzle of the structure, composition, and density of the solar atmosphere was fitted into place.

In 1877, Young was lured away from Dartmouth by the offer of more equipment and research time at the College of New Jersey, now Princeton University. Not only was he a great researcher, but a revered teacher and acclaimed writer. His textbooks were the standard of his day.

The Legacy of Eclipses

Throughout the final three decades of the nineteenth century, Janssen, Lockyer, and Young led expeditions to the major total eclipses and, weather permitting, always contributed useful data and often new discoveries.

Many scientists had noticed that the corona changes its appearance from one eclipse to the next. But it was Jules Janssen who first spotted a pattern to those variations. He compared the coronas of the 1871 and 1878 eclipses and concluded that the shape of the corona varies according to the sunspot cycle. In 1871, the Sun was near sunspot maximum and the corona was round. In 1878, near sunspot minimum, the corona was more concentrated at the Sun's equator.

As the twentieth century began, solar eclipses were still the principal means of gathering information about the workings of the Sun. Every total eclipse over land was attended by scientists willing to travel great distances, endure hostile climates, and risk complete failure because of clouds for a few minutes' view of the corona—vital for the systematic study of the Sun launched more than half a century earlier by Francis Baily in his report on the annular eclipse of 1836.

One could see the Sun best when it was obscured.

Bolivia (November 12, 1966); dot to right is Mercury

Mexico (March 7, 1970)

India (February 16, 1980)

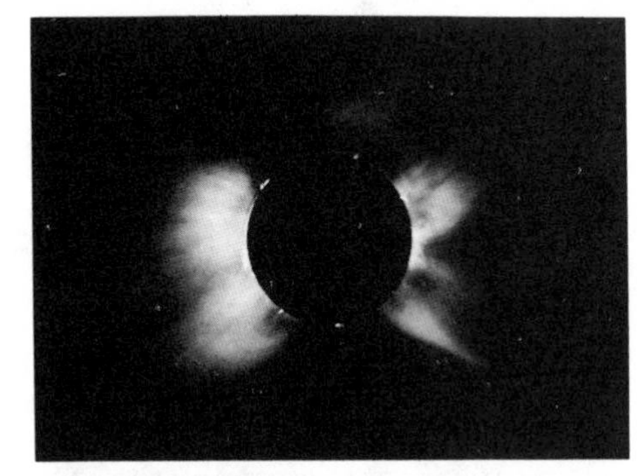

Siberia (July 31, 1981)

Java (June 11, 1983)

Philippines (March 18, 1988)

The shape of the corona changes with sunspot activity. Near sunspot minimum the corona is elongated at the Sun's equator. Near maximum, the corona is more symmetrical. These photographs were taken with a radial density filter that captures faint detail in the outer corona without overexposing the inner corona—similar to what the eye sees. The corona never appears the same twice. *High Altitude Observatory/NCAR*

Oh leave the Wise our measures to collate
One thing at least is certain, LIGHT has WEIGHT
One thing is certain, and the rest debate—
Light-rays, when near the Sun, DO NOT GO
STRAIGHT.

Arthur S. Eddington (1920)

8

A Magic Shadow Show

Of all the lessons that scientists learned from eclipses, the most profound and momentous was the confirmation of Einstein's general theory of relativity by the eclipse of May 29, 1919.

The Birth of Relativity

In 1905, an obscure Swiss patent examiner third class named Albert Einstein published three articles in the same issue of the leading German scientific journal *Annalen der Physik* that utterly changed the course of physics. One proved the existence of atoms. The second laid the cornerstone for quantum mechanics. The third was a revolutionary view of space and time known as the special theory of relativity. It required only high-school algebra, yet its implications were so profound that this theory baffled many of the leading scientists of the day.

In the decade that followed the publication of the special theory of relativity, Einstein labored mightily to expand his concept to accelerated systems. In 1907, he formulated his principle of equivalence: there is no way for a participant to distinguish between a gravitational system and an accelerated system. For an observer in a closed compartment, does a ball fall to the floor by gravity because the compart-

Einstein's principle of equivalence. *Drawn by Tim Phelps*

ment is resting on a planet or does the ball fall because the compartment is accelerating toward the ball? In both cases, the ball falls to the floor. In both cases, the observer feels weight. It is impossible to tell whether that weight is from gravity or acceleration.

Einstein realized from his principle of equivalence that relativity required gravity to bend light rays, much as it bends the paths of particles. In a closed accelerating compartment, a light on one wall is aimed directly at the opposite wall, across the line of motion. The beam travels at a finite speed, so in the time it takes to traverse the compartment, the opposite wall has moved upward. For an observer in the compartment, the light beam has struck the opposite wall below where it was aimed. The observer concludes that light has been bent.

What about that same experiment performed in a compartment at rest on a planet? According to Einstein's principle of equivalence, there can be no difference between phenomena measured in the two compartments. Therefore, for the observer in the gravitational environment, light must be bent by gravity. But the effect is very small. It takes a lot of mass to bend light enough to be measured. In 1911, Ein-

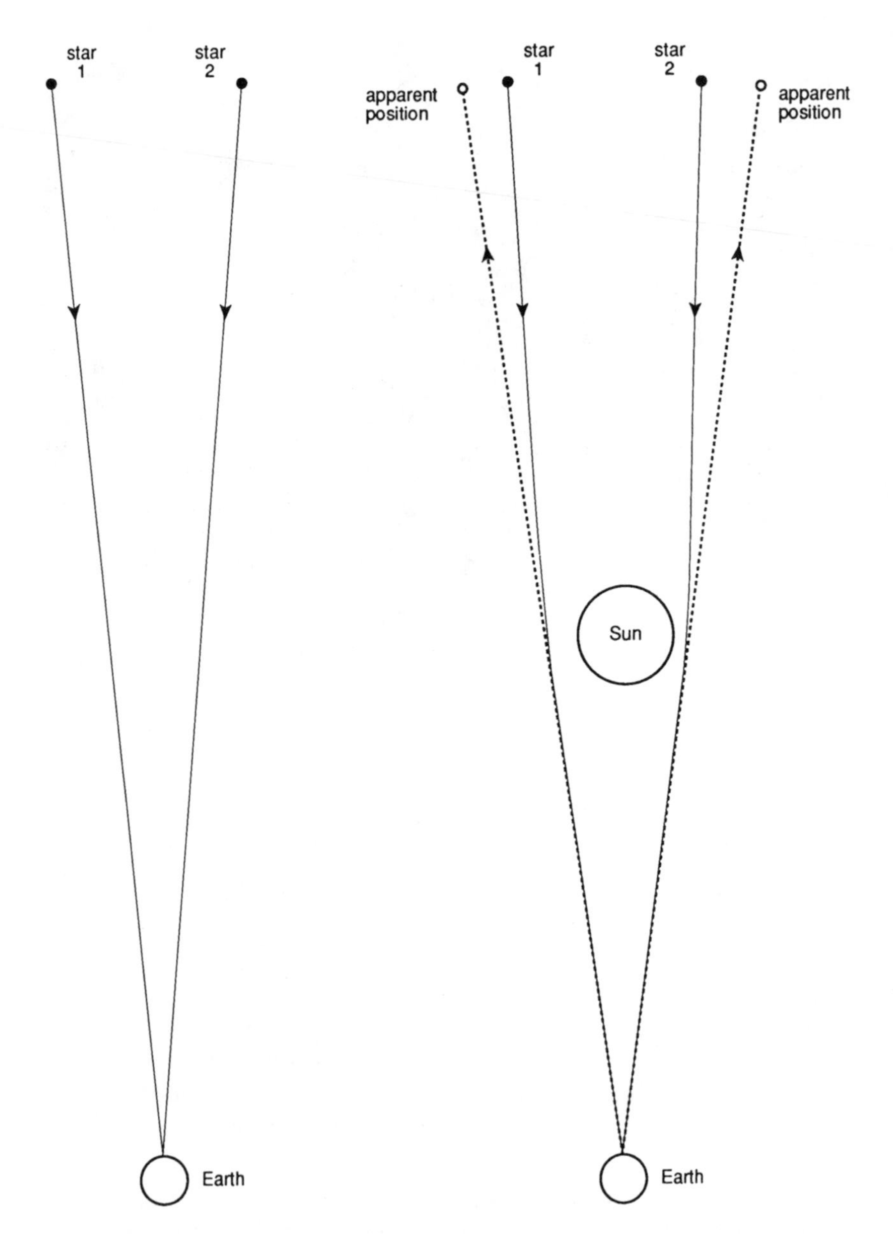

In a perfect vacuum with no gravity, light from distant stars travels in a straight line, as illustrated at left. The gravitational field of the Sun bends light (shown at right), making the stars appear slightly farther apart than they actually are.

stein realized that this peculiar idea might be tested during a total solar eclipse.

The Sun is sufficiently massive that light from distant stars passing close to its surface would be deflected just enough to be measurable. This bending of the light from stars causes them to appear displaced slightly outward from the Sun. The displacement could be recognized by comparing a photograph of the star field around the Sun with a photograph of the same star field when the Sun was not present.

When the Sun is visible, however, the stars are much too faint to be seen. If only the Sun's brightness could be shut off for just a few minutes—as happens during a total solar eclipse! Using his special theory of relativity, Einstein initially predicted that starlight just grazing the Sun would be bent 0.87 second of arc.[1]

The Search for Proof

A talented young scientist named Erwin Freundlich was the first to attempt to test this aspect of relativity.[2] Fascinated by the theory, he obtained solar eclipse photographs from observatories around the world that might show the displacement of stars. But photographs of previous eclipses were not adequate for the purpose.[3] The theory would have to be tested at a future eclipse, such as the one that would occur in southern Russia on August 21, 1914.[4] Freundlich was an assistant at the Royal Observatory in Berlin and tried to interest his colleagues in mounting a scientific expedition to test the theory. His superiors, however, were uninterested. Freundlich was allowed to go if he took unpaid leave and raised his own money. With youthful enthusiasm, he made his plans and informed Einstein of his intentions.

Einstein had just moved from Switzerland to Berlin to take a distinguished position created for him at the Kaiser Wilhelm Institute, the most prestigious research center in the world's science capital. Even among the giants there, Einstein stood out. Physicist Rudolf Ladenburg recalled, "There were two kinds of physicists in Berlin: on the one hand was Einstein, and on the other all the rest."

It was at that time, the spring of 1914, that Einstein was becoming increasingly withdrawn and oblivious to conventions of social behavior. He was confident in his new general theory of relativity but was deeply engrossed in its final formulation and its implications. Freundlich's wife Käte told of inviting Einstein to dinner one evening. At the conclusion of the meal, as the two scientists talked, Einstein suddenly pushed back his plate, took out his pen, and began to cover their prized tablecloth with equations. Years later, Mrs. Freundlich lamented, "Had I kept it unwashed as my husband told me, it would be worth a fortune."[5]

That summer Freundlich took his scientific equipment and headed for the Crimea and the eclipse. On August 1 Germany declared war on Russia, commencing World War I. Freundlich was a German behind Russian lines. He and his team members were arrested and their equipment impounded. Within a month, Freundlich and crew were exchanged for high-ranking Russian officers, but they had missed the eclipse.[6]

Einstein deplored the war and German militarism, an attitude that drew the ire of many of his colleagues in Berlin. He ignored the hostility and concentrated all his energies on his research. In 1915 Einstein announced the completion of his general theory of relativity, a radically new theory of gravitation. It is perhaps the most prodigious work ever accomplished by a human being. Not only were its implications profound, but the mathematics needed to understand it were formidable. "Compared with [the general theory of relativity]," said Einstein, "the original [special] theory of relativity is child's play."[7]

Einstein offered three tests of his theory. The first was the peculiar motion of Mercury. The entire orbit of Mercury was turning (precessing) more than Newton's law of universal gravitation could explain. It was a tiny but measurable amount: 43 seconds of arc a *century*. This unexplained advance in Mercury's *perihelion*, its closest point to the Sun, had given rise to a suspicion that one or more planets lay between Mercury and the Sun. The suspicion was so strong that this suspected planet, scorched by the nearby Sun, had already been given a name: Vulcan, after the Roman god of fire. Many observers had tried to spot it as a tiny dot passing across the face of the Sun, and some even claimed success. But when an orbit for the planet was calculated and its next passage across the face of the Sun was due, the planet never kept the appointment. It did not exist. But it was not until Einstein formulated the general theory of relativity that the 43 seconds per century anomaly could be explained.

Einstein was more than just pleased when he realized that his

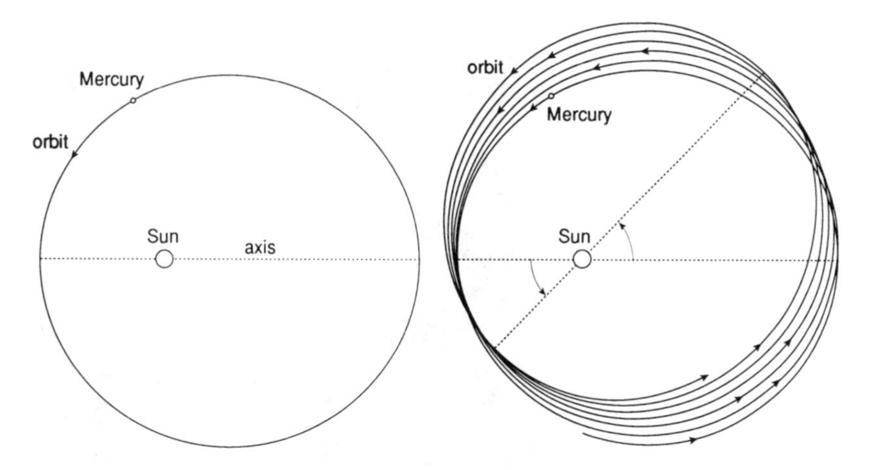

Precession of Mercury's orbital axis: *Left,* according to Newtonian theory, a planet orbiting the Sun follows a fixed elliptical path; *right,* Einstein's general theory of relativity predicted that the axis of the ellipse would gradually rotate.

theory could account for this discrepancy in the motion of Mercury. "I was beside myself with ecstasy," he wrote.[8] This explanation of a perplexing problem gave general relativity high credibility. But the power of the general theory would be even more evident if it could predict something never before contemplated or detected.

Einstein offered two such predictions: that starlight passing close to the Sun would be bent, and that light leaving a massive object would have its wavelengths extended so that the light would be redder. This *gravitational redshift* was so small an effect that it could not be detected in the Sun with the equipment available at the time, so this proof of general relativity had to wait many years. It was finally detected in 1959 by Robert V. Pound and Glen A. Rebka, Jr., using the recently discovered Mössbauer effect in which the gamma rays emitted by atomic nuclei serve as the most precise of clocks. One of these atomic clocks was placed in the basement of a building and moved up and down so that its depth in the Earth's gravitational field varied minutely. The deeper in the basement the clock was, the longer the wavelength of its radiation. It had taken 45 years, but the gravitational redshift predicted by Einstein had at last been confirmed.[9]

In contrast, the gravitational deflection of starlight predicted by Einstein could be tested at most total eclipses of the Sun. Between 1911 and 1915, Einstein revised his calculation, using his new general

Albert Einstein in 1922.
Burndy Library & AIP Niels Bohr Library

theory of relativity. He found the deflection to be twice the initially assigned value. Starlight passing near the Sun would be bent 1.75 arc seconds. (Einstein and the world were fortunate that his initial prediction was not tested before it was revised; otherwise his later figure, although rigorously honest, might have seemed to be a manipulation to make the numbers come out right. There would have been far less drama in the confirmation of relativity, and the theory might be only vaguely known to the general public.)

In 1916 Einstein published his complete general theory of relativity. But the world hardly noticed. For two years, nations had been locked in the First World War. Feelings against Germans ran strong in France and Great Britain, just as the Germans hated the French and British. Einstein sent his paper to a friend, scientist Willem de Sitter in The Netherlands. De Sitter, in turn, forwarded the paper to Arthur Eddington in England. At the age of thirty-four, Eddington was already famous for his pioneering work in how a star emits energy. Eddington instantly recognized the significance of Einstein's discoveries and was deeply impressed by its intellectual beauty. He shared the paper with other scientists in Britain.

Arthur S. Eddington. *AIP Niels Bohr Library, gift of S. Chandrasekhar*

The 1919 Test

Frank Dyson, the astronomer royal of England, began planning for a British solar eclipse expedition in 1919 to test relativity. It was the perfect eclipse for the purpose because the Sun would be standing in front of the Hyades, a nearby star cluster, so there would be a number of stars around the eclipsed Sun bright enough for a telescope to see. The timing could hardly have been worse. In 1917 Britain was in the midst of a terrible war whose issue was still very much in doubt. Yet somehow Dyson managed to persuade the government to fund the expedition, despite the fact that its purpose was to test and probably confirm the theory of a scientist who lived in Germany, the leader of the hostile powers.

Meanwhile, American astronomers had an earlier opportunity to verify or disprove relativity. In the final months of World War I, a total eclipse passed diagonally across the United States from the state of Washington to Florida. On June 8, 1918, a Lick Observatory team led by William Wallace Campbell and Heber D. Curtis, observing from Goldendale, Washington, pointed their instruments skyward to render a verdict on relativity. The weather was mostly cloudy, but the Sun broke through for three minutes during totality. Measuring and interpreting the plates had to wait several months, however, until comparison pictures could be taken of the same region of the sky at night, without the Sun in the way. By this time, Curtis was in Washington, D.C., working on military technology for the government. Curtis returned to Lick in May 1919 and the results were announced in June. The star images were not as pointlike as desired. Curtis could detect no deflection of starlight. By this time, word was spreading about the findings of the British expeditions, and the Lick paper was never published.[10]

Four months after the armistice, two British scientific teams were poised for departure. Andrew C. D. Crommelin and Charles R. Davidson, heading one party, were to set sail for Sobral, about 50 miles (80 km) inland in northeastern Brazil. Arthur Eddington, Edwin T. Cottingham, and their team were headed for Principe, a Portuguese island about 120 miles (200 km) off the west coast of Africa in the Gulf of Guinea. On the final day before sailing, the four team leaders met with Dyson for a final briefing. Eddington was extremely enthusiastic and confident that Einstein was right. A deflection of 1.75 arc seconds would confirm relativity. Half that amount—a deflection of starlight by 0.87 arc seconds—would reconfirm Newtonian physics.[11]

"What will it mean," asked Cottingham, "if we get double the Einstein deflection?" "Then," said Dyson, "Eddington will go mad and you will have to come home alone!"

On Principe worrisome weather conditions greeted Eddington and Cottingham. Every day was cloudy. However, May was the beginning of the dry season and no rain fell—until the morning of the eclipse. The fateful day dawned overcast, and heavy rain poured down. The thunderstorm moved on about noon, but the cloud cover remained. The Sun finally peeped through 18 minutes before the eclipse became total but continued to play peekaboo with the clouds. Said Eddington: "I did not see the eclipse, being too busy changing plates, except for one glance to make sure it had begun and another half-way through to see how much cloud there was."

Eddington had much cause for worry. He was not interested in prominences or the corona. He needed to see clearly the region around the Sun. With great care, the plates were developed one at a time, only two a night. Eddington spent all day measuring the plates. Clouds had interfered with the view, but five stars were visible on two plates. When he had finished measuring the first usable plate, Eddington turned to his colleague and said, "Cottingham, you won't have to go home alone."

Eddington measured the displacement in the stars' positions, extrapolated to the limb of the Sun, to be between 1.44 and 1.94 arc seconds, for a mean value of 1.61 ± 0.30 arc seconds. The deflection agreed closely with Einstein's prediction. In later years, Eddington referred to this occasion as the greatest moment of his life.[12]

The expedition to Sobral had equally threatening weather but there was a clear view of totality except for some thin fleeting clouds in the middle of the event. Crommelin and Davidson stayed in Brazil until July to take reference pictures of the star field without the presence of the Sun and then brought all their photographic plates back to Britain before measuring them. They found that their largest telescope had failed because heat caused a slight change in focus, which spoiled the pinpoint images of the stars. But the other instrument had worked well, and its plates also supported Einstein's prediction. Their mean value was 1.98 ± 0.12 arc seconds.

It was September and no news about eclipse results had been released. Einstein was curious and inquired of friends. Hendrik Antoon Lorentz, the Dutch physicist, used his British contacts to gather news and telegraphed Einstein: "Eddington found star displacement at rim of Sun . . . ."[13] He also announced the favorable results at a scientific meeting in Amsterdam on October 25, with Einstein in attendance. But no reporters were present, and no word of the discovery was published.

Finally, on November 6, 1919, the Royal Society and the Royal Astronomical Society held a joint meeting to hear the results of the eclipse expeditions. The hall was crowded with observers, aware that an age was ending. From the back wall, a large portrait of Newton looked down on the proceedings. Joseph John Thomson, president of the Royal Society and discoverer of the electron, chaired the meeting, praising Einstein's work as "one of the highest achievements of human thought." He called upon Frank Dyson to summarize the results and introduce the reports of the eclipse team leaders. The astronomer royal concluded his presentation by saying: "After careful study of the plates I am prepared to say that there can be no doubt

Albert Einstein and Arthur S. Eddington in Eddington's garden, 1930. *Royal Greenwich Observatory*

that they confirm Einstein's prediction. A very definite result has been obtained that light is deflected in accordance with Einstein's law of gravitation."[14]

Einstein awoke in Berlin on the morning of November 7, 1919, to find himself world famous. Hordes of reporters and photographers converged on his house. He genuinely did not like this attention, but he found a way to turn the disturbance to the benefit of others. He told the reporters about the starving children in Vienna. If they wanted to take his picture, they first had to make a contribution to help those children. Suddenly Einstein was a celebrity.

William Wallace Campbell about 1914. *Mary Lea Shane Archives of the Lick Observatory*

More Tests

Expeditions from many nations set off to retest relativity by measuring the deflection of starlight during the eclipse of September 21, 1922. The shadow passed across Somalia, the Indian Ocean, and Australia. Most of the teams had bad luck; the weather failed them. But the successful American delegation, a Lick Observatory team again headed by Campbell, observed the eclipse from Wallal on the northwest coast of Australia. After the Lick Observatory disappointments in 1914 and 1918, this expedition was an all-out effort, with the best equipment and lots of rehearsal. The results had to wait, though, until comparison photographs had been taken and until Campbell and Robert C. Trumpler had measured the plates with utmost care. In April 1923, seven months after the eclipse, Campbell announced: "The agreement with Einstein's prediction from the theory of relativity . . . is as close as the most ardent proponent of that theory could hope for."[15] They measured the displacement at 1.72 ± 0.11 arc seconds.

When Campbell was asked for his personal reaction to this new

Heber D. Curtis about 1913. *Mary Lea Shane Archives of the Lick Observatory*

confirmation of relativity, he replied: "I hoped it would not be true."[16] Campbell was not the only one reluctant to accept such a new and difficult concept. There were a number of scientists who did not share Eddington's delight in Einstein's victory.

For the next half century, most of the Sun's rendezvous with the Moon were attended by scientists carrying instruments to remeasure the deflection of starlight. A wide range of conflicting values resulted. For the eclipse of June 30, 1973, a definitive measurement was attempted. Once again, the eclipse was conducive for study: the second longest totality of the twentieth century. A combined University of Texas and Princeton University team journeyed into the Sahara to watch the eclipse from an oasis in Mauritania where totality would last 6 minutes 18 seconds. They took special precautions about temperature variations, emulsion creepage, and other factors. The precautions were almost for naught. Eclipse day brought a cruel surprise.

The 1922 Lick Observatory eclipse expedition unloads equipment at Wallal, Australia. *Mary Lea Shane Archives of the Lick Observatory*

Makeshift mount for the principal telescope used by the 1922 Lick Observatory team to test relativity. *Mary Lea Shane Archives of the Lick Observatory*

Women's work at the 1922 expedition. Note the mosquito netting. *Mary Lea Shane Archives of the Lick Observatory*

Before the Moon blocked the Sun's light, a dust storm provided its own eclipse, subsiding only a few minutes before totality, but leaving the atmosphere choked with dust that blocked out 82 percent of the light.[17] Still, the astronomers recorded 150 measurable images and found a deflection of 1.66 ± 0.18 arc seconds, well in accord with Einstein's prediction.

Even as that attempt was made, a new method of testing relativity by the deflection of radiation around the Sun was emerging. Radio astronomers were using widely separated radio telescopes to measure positions with an accuracy greater than could be obtained with optical telescopes. Quasars had been discovered and nearly all astronomers interpreted them to be the most distant objects in the universe—so distant that they were essentially fixed markers. Their angular distance from one another provided a new coordinate system against which the positions of all other objects could be referred. Because some of the quasars lay near the ecliptic, the Sun's apparent path through the heavens in the course of a year, the Sun's gravity would annually deflect the quasars' radiation, making them seem to shift

Solar eclipse as photographed at Wallal, Australia (September 21, 1922). *Mary Lea Shane Archives of the Lick Observatory*

slightly in position. Only this time, the radiation would be radio waves rather than visible light. No eclipse was necessary because radio telescopes do not require darkness. In 1974 and 1975, Edward B. Fomalont and Richard A. Sramek used a 22-mile (35-km) separation between radio telescopes to measure the deflection of light at the limb of the Sun as 1.761 ± 0.016 arc seconds, a result that not only confirmed relativity, but also favored Einsteinian relativity over slightly variant relativity formulations by other scientists.[18]

Solar eclipses may no longer be the most accurate way of determining the relativistic deflection of starlight, but when the general theory of relativity needed proof from a phenomenon predicted but never before observed, it was solar eclipses that provided the first and most dramatic demonstration of Einstein's masterpiece and brought relativity and Einstein to the attention of the entire world.

I have a little shadow that goes in and out with me,
And what can be the use of him is more than I can
see.

Robert Louis Stevenson (1885)

9 Modern Scientific Uses for Eclipses

When scientists first turned their attention to solar eclipses, they put them to use to clock the motions of the Moon around the Earth and the Earth around the Sun. By trying to predict the exact time and location of the path of a solar eclipse, astronomers could take note of their errors and refine their knowledge of the orbits of the Earth and Moon. This work was pioneered by Edmond Halley in 1715 for a total eclipse crossing southern England.

More than a century later, in the early days of astrophysics, a second use for total eclipses emerged. The eclipse of 1842, carving a path across southern France and northern Italy, gave European scientists a front-row seat to see for themselves the occasionally reported effects surrounding totality: the corona, prominences, and chromosphere. They were awed and realized that the Sun, by covering its face, was revealing physical aspects of itself that were not visible at any other time. By studying the atmosphere of the Sun during total eclipses and the visible surface of the Sun at all other times, scientists began to perceive how hot the interior of the Sun must be.

Then, unexpectedly, early in the twentieth century, there arose a third great scientific use for solar eclipses: to confirm or deny the peculiar new theory of gravity and the structure of the cosmos offered by Albert Einstein. The first affirmative answer came from an eclipse in 1919, and additional data from subsequent eclipses poured in for the next half century and more.

Meanwhile, the original scientific uses for total eclipses waned. The U.S. Naval Observatory no longer refines the orbits of the Moon or Earth from eclipse timings. Equipment and techniques developed since 1868 allow astronomers to study the prominences of the Sun independent of eclipses. In 1930, Bernard Lyot of France invented the coronagraph, a telescope with a special optical system to create an artificial eclipse so that the brighter regions of the corona can be studied without waiting for a precious few moments of eclipse totality in some remote corner of the world. And the relativistic bending of starlight can now be checked more precisely by using radio waves, which can be received during broad daylight, without waiting for an eclipse.

Have eclipses been used to exhaustion by scientists and then abandoned to the care of esthetes and amateur astronomers?

Dogging the Sun's Diameter

David Dunham is not about to consign eclipses to the historical archives of scientific relics. He and his colleagues Joan Dunham, Alan Fiala, Paul Maley, amateur astronomers David Herald of Australia and Hans Bode of Germany, and other members of the International Occultation Timing Association (IOTA) journey to eclipses around the world for the express purpose of almost missing them—that is, missing what almost everyone else is most anxious to see: the longest possible duration of totality. You will never find Dunham and his colleagues along the center line. Instead, they and any friends or local people they can recruit position themselves across the northern and southern limits of the eclipse path so that they witness just a few moments of totality, or none at all.

From their positions at the edges of totality, the top or bottom of the Moon just briefly covers the full face of the Sun. The corona no sooner appears than it fades away as the Sun reemerges. This is precisely the information Dunham seeks. He and his co-workers are trying to measure minute changes in the diameter of the Sun and have enlisted the Moon for their service. The size and distance of the Moon are well known. The distance of the Sun is known to great accuracy. Therefore, by measuring the size of the Moon's shadow, they can derive the size of the body responsible for that shadow: the Sun.

By stretching a team of observers perpendicular to the expected edge of the eclipse path, typically from 0.5 mile (0.8 km) outside to 1.5 miles (2.4 km) inside, they can determine where the actual edge of totality passes to within a hundred meters. This translates into the ability to measure the angular radius of the Sun to an accuracy of 0.04 arc second or about 20 miles (30 km). The Sun in 1983 was, according to their measurements, about 0.4 to 0.5 arc second larger than it was in 1979, which means that the Sun had increased in radius by about 180 miles (290 km). However, the Sun seemed to be 0.5 arc second smaller in 1979 than it was in 1715 or even 1925, a decrease in the solar radius by about 230 miles (375 km).[1]

It is a little surprising to think that our Sun might be expanding, shrinking, or pulsating. But, then, it was a shock to people almost four centuries ago when Galileo used his telescope on the Sun, a celestial body thought to be "pure" and immune to change, and found spots on the Sun that appeared, changed, disappeared, and showed that the Sun was rotating. Over the centuries, scientists recognized more and more changes on the Sun: the shape of the corona, prominences, flares, the sunspot cycle. What is most surprising is not that the Sun changes but that changes on the Sun are not reflected more dramatically (and catastrophically) in short-term climate changes on Earth.[2]

Do the Dunhams, Fiala, and their colleagues regret missing the "main attraction" of a total eclipse? No, says Dunham, he actually prefers the view from the edge. He sacrifices a long look at the corona and the eerie twilight at the center of the eclipse path, but he gains a maximum duration view of Baily's Beads, the chromosphere, and the shadow bands, which he sees without special effort at virtually every eclipse. Dunham has observed seven eclipses in this fashion and, funds permitting, he plans to attend many more.

If this variation in the diameter of the Sun is real, and he and his colleagues think it is, they suspect that it may be a cyclic pulsation tied to the sunspot cycle of approximately 11 years. Why, Dunham does not know. He refers theoretical questions to others, such as Sabatino Sofia. Dunham concentrates on refining his equipment and techniques to hold experimental error to low enough levels so that his data will have clear meaning.

The task is simple in theory but very complicated in practice. The raw data—what was seen from each precisely marked viewing position at precisely what time—are just the start of the project. To compute the angular size of the Sun to see if it has changed, one must take into consideration a host of factors, the most complex of which is the landscape of the Moon. Because of the mountains and valleys on the Moon, its limb is jagged, producing Baily's Beads as totality nears and the only light from the Sun's photosphere that reaches the Earth is passing through the deepest valleys. But because of the Earth's angle of viewing, the features along the Moon's edge appear to change constantly. Depending on our latitude, we see a little over or under the Moon's poles. From the east or west, we see the polar features at slightly different angles. This change of perspective with each eclipse and each position along the eclipse path causes the apparent positions and angular heights of features to change minutely. In addition, the Moon actually wobbles slightly due to the tidal effects of the Earth and Sun, adding in a complex way to this libration effect.

For the radius of the Sun to be measured accurately using the Moon's shadow during an eclipse, the exact height of the north and south limb features of the Moon as they appear during each specific eclipse must be calculated to high accuracy, otherwise the fraction of a mile uncertainty in the bead-creating diameter of the Moon washes out the accuracy in the measurement of the Sun's radius. Critics of solar radius variation claims say that practitioners are overestimating the accuracy of their measurements.[3]

If the Sun actually does pulsate slightly in the course of a sunspot, and hence magnetic, cycle, it will probably take another decade or two to prove the point—a long-term project that brings to mind the cartoon of the turtle family, with the young turtle a few steps distant from his parents. "Don't look now," says Mother Turtle, "but I think Junior is running away again. The next few weeks will tell."[4]

Across the Spectrum

Dunham and IOTA members are by no means the only astronomers who still see important scientific value in solar eclipses. In fact, solar eclipses provide so many unique opportunities to study the Sun's atmosphere that this chapter can present only a small sampling of the research in progress.

Spectroheliographs and coronagraphs allow photographic inspection of many aspects of the Sun's chromosphere, prominences, and lower corona outside of an eclipse. But observing in special wavelengths or by placing an occulting disk at the focal plane to block the face of the Sun cannot compensate for the turbulence in the Earth's atmosphere that causes the sunlight to be refracted so that it comes in from slightly different angles. This deflection of the light blurs the

---

The Corona, Eclipses, and Modern Solar Research

by Jay M. Pasachoff

The disk of the Sun is visible from the ground every clear day, but the solar corona is most clearly visible from the ground only during total eclipses. Scientists are interested in the corona for four major reasons.

First, we learn about how the Sun works. All the energy that leaves the Sun passes through the corona. Why the temperature of the corona is as high as 2 million degrees Celsius is still a major question, with theories of coronal heating via magnetic fields currently dominating.

Second, we study the solar corona to learn more about the Earth. The corona continually expands into space in the form of the solar wind, which envelops our planet. Thus the Sun causes many effects on Earth: the aurorae, short-wave and CB communication blackouts, and surges on power lines that lead to outages. Changes in solar radiation may affect the Earth's climate on a time scale short enough to detect.

Third, the Sun is a rather typical star, so by studying the Sun we gain detailed closeup information that we can apply to the distant stars that are mere pinpoints of light in our largest telescopes. The Sun is therefore a key to understanding the universe.

Fourth, the Sun is a celestial laboratory where we learn basic laws of physics. The conditions on the Sun are often not duplicable on Earth. For example, the density of the solar corona is so low that it would be considered a fantastic vacuum in a laboratory on Earth. The corona is a hot plasma that the solar magnetic field directs into the beautiful streamers seen during eclipses. Thus the corona reveals the behavior of hot gas held in a magnetic field. This knowledge is useful to magnetic fusion research, which may one day provide energy on Earth. In the laboratory, we have trouble holding plasma together long enough for protons and deuterons to come sufficiently close together to fuse, releasing energy according to Einstein's formula $E = mc^2$. Positively charged particles repel each other strongly. In a hot gas, particle velocities are so high that protons and deuterons approach each other closely despite their mutual repulsion. But no material container can hold such hot gas, so magnetic fields are used to constrain it. Astronomical studies of the corona help physicists and engineers with this important problem.

---

image enough to suppress viewing of fine detail in the corona and prominences altogether, leaving only the most conspicuous solar atmospheric features visible.

Jack Zirker and his co-workers would like to use modern advances in instrumentation to record the corona and the promi-

---

The corona is so important to science that no method of observing it should be ignored. Solar telescopes in space, such as those of the Orbiting Solar Observatories, Skylab, the Solar Maximum Mission, and Spacelab 2 (aboard the Space Shuttle), have observed the Sun in parts of the spectrum that do not reach the ground and have provided magnificent new insights. But to limit the scattering of sunlight, space-borne coronagraphs to date have had to block out the inner corona. And ground-based observations of the corona outside of eclipses cannot see the corona as far from the Sun as can be seen at a total eclipse. Thus, eclipse observations are essential supplements to ground- and space-based coronal observations if the full picture is to be grasped.

The traditional advantages of eclipses over space observations for solar research are flexibility and cost. New observations and instruments can be incorporated into an eclipse expedition on short notice. State-of-the-art and bulky equipment can be transported to eclipse sites far more cheaply than to space. Eclipse equipment does not have to meet launch standards of sturdiness. Further, eclipse instruments can be mounted on steady bases, with the Earth as a platform, and can be adjusted at the last minute by qualified scientists. Because of spacecraft jitter and limitations on the size of orbiting telescopes, no spacecraft has yet been able to study details on the Sun as fine as those seen from Earth.

Besides, solar spacecraft are few and sometimes far between. The Solar Maximum Mission satellite burned up in the Earth's atmosphere in 1989. The Solar Optical Telescope, now downsized to the High Resolution Solar Observatory, is presently on hold. And Spacelab will not fly with more solar instruments for many years. So the hope of solar physicists to have high resolution in space will not happen soon.

The Skylab solar telescopes and the Solar Maximum Mission cost hundreds of millions of dollars each. For less than one-tenth of one percent of these costs, a well-equipped modern ground-based eclipse expedition can be mounted. Even allowing for some of the eclipses to be clouded out, eclipses are a cost-efficient way of doing astronomy research.

---

nences in high resolution. By taking numerous short-exposure pictures, they hope to catch small but rapid changes in the corona or prominences that have never been noticed before. They do not know quite what to expect, but an eclipse such as that in 1991 offers the best opportunity for such an experiment. Astronomers are used to dragging telescopes thousands of miles to eclipse sites. Yet only the smaller telescopes are portable. It would be nice if a total eclipse came to a large observatory for a change.

One does. The center line of the 1991 eclipse climbs over Mauna Kea, Hawaii's largest volcanic peak, and passes within 1.5 miles (2.4 km) of the largest assemblage of telescopes in the world, offering more than four minutes of totality. Zirker and his colleagues Bill Livingston and Barry LaBonte will be using one of the large telescopes on Mauna Kea, the 88-inch (2.2-m) University of Hawaii reflector, to record the corona with very short exposures, hoping to see fine detail never before detected.

Charles Lindsey and Eric Becklin use eclipses to study the infrared radiation emitted from the Sun's chromosphere. They have refined earlier innovations of Alan Clark and John Beckman, pioneers in eclipse observations from aircraft. Because much of the infrared spectrum is absorbed by water vapor in the Earth's atmosphere, it is impossible to study wavelengths from 35 to 300 microns from sea level. So, during the total eclipses of 1981 and 1988, Becklin and Lindsey's team flew above most of the atmosphere in NASA's Kuiper Airborne Observatory, a military cargo jet converted to carry a 36-inch (0.9-m) telescope. By measuring the rate of disappearance of infrared radiation as the Moon covered the Sun, they were able to determine a profile of the Sun's chromosphere. The profile gives the extent and temperature of the chromosphere at selected infrared wavelengths from 30 to 800 microns. This infrared profile provides evidence of dense rough structure deeper in the chromosphere than expected, and enables theorists to construct more realistic models of how the chromosphere works.[5]

Thanks to exceptionally dry conditions atop high mountains, infrared wavelengths can be recorded there without the use of aircraft. New, very large infrared telescopes have recently been installed on Mauna Kea that enable astronomers to make observations in extreme infrared wavelengths far beyond the capabilities of the small telescopes aboard airplanes. Lindsey and his colleagues from the United States,

Canada, and Great Britain hope to observe the 1991 eclipse with the 49-foot (15-m) James Clerk Maxwell Telescope, operated on Mauna Kea by the United Kingdom, the Netherlands, and Canada.

During solar eclipses long ago, the Chinese beat drums, clanged cymbals, and fired off skyrockets to scare away the dragon or dog that was eating the Sun. In modern times, Gary Rottman and Frank Orrall still fire off rockets during solar eclipses, but their intent is discovery.

They use NASA suborbital rockets, such as the Black Brant, to carry their ultraviolet camera and spectrometer to an altitude of about 200 miles (325 km), where extremely short ultraviolet wavelengths from the Sun's corona can be recorded before they are absorbed in the Earth's atmosphere. Actually, Rottman and Orrall do not need an eclipse of the Sun in order to gather these data on the corona. The high temperature of the corona gives it a brightness at very short wavelengths that overwhelms the brightness of the comparatively cool photosphere. But total eclipses are the best time for rocket flights because the state of the corona they record in their five minutes above the Earth's atmosphere can be compared with the corona as seen from the ground in visible and radio wavelengths. Each wavelength provides its own revelations about the corona and, studied together, they provide powerful tests of the theories that attempt to explain peculiar features of the corona: its chemical abundances, its density irregularities (clumpiness), how it is heated to exceptionally high temperatures, and the mechanism by which the Sun drives the solar wind.

The Chaos of the Corona

The atmosphere of Earth is controlled by three primary forces: gravity, pressure variations, and the Earth's rotation. The Sun's atmosphere is shaped not only by gravity, pressure, and rotation, but by a fourth force: magnetism. A total eclipse offers the best chance to see how magnetic fields sculpt the atmosphere of the Sun. By revealing the corona, totality also provides a chance to study a high temperature plasma under conditions that cannot be duplicated on Earth: extremely hot ionized gases at extremely low density. A total eclipse offers as well an opportunity to see how the corona changes with time and to what degree these fluctuations correlate with magnetic activ-

Loop prominence. *NSO/NOAO*

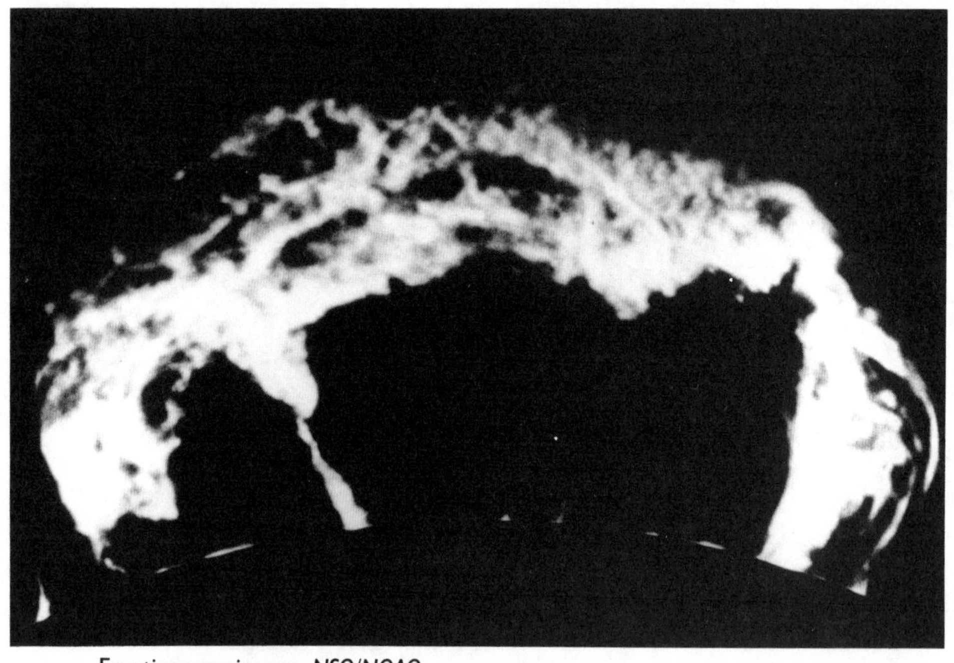

Eruptive prominence. *NSO/NOAO*

ity. Richard R. Fisher is one of a team of scientists who like to use eclipses to record the temperature and structure of the corona with far more precision than observations of the Sun outside of eclipses provide.

Another astronomer still using eclipses to probe the Sun is Jay Pasachoff. He is eager to understand how the corona is heated to temperatures more than 300 times hotter than the photosphere. The energy of the Sun is created at its core, where the temperature is about 27 million °F (15 million °C). By radiation and then convection the photons of light energy make their way to the surface, gradually cooling enroute, so that the Sun's photosphere glows with a temperature of only about 10,000 °F (5,500 °C). The base of the chromosphere, the lower atmosphere of the Sun, is cooler still, about 7,200 °F (4,000 °C). But the temperature in the corona exceeds one million degrees. Temperature is a measure of the random motion of atoms and subatomic particles, and the particles in the corona are indeed traveling at great speed. But the corona has a very low density, so the total amount of heat in the corona is quite small.

What causes the temperature in the corona to be so much higher than that of the visible surface of the Sun? These surprising temperatures seem to be caused by magnetic fields. The same magnetic fields that cause sunspots in the photosphere seem to pervade the corona as well. They are responsible for streamers and other coronal features; they also shape the prominences and filaments. Most scientists now favor the theory that the corona is heated to its extreme temperatures by resistive dissipation of electric currents associated with the magnetic fields, a problem on which theoretical work by Eugene Parker has shed considerable light in recent years.[6]

Another possibility is that magnetic waves (Alfvén waves), excited by motion in the convective region beneath the photosphere, propagate energy into the corona where it is then dissipated into heat. Such waves might manifest themselves as local oscillations in coronal brightness with periods in the range of 1 to 10 seconds. Pasachoff is interested in seeing these tiny but rapid oscillations, if they exist. He uses every accessible total eclipse to search for these oscillations that he believes heat the corona of his favorite star.

Now eclipses are elusive and provoking things . . . visiting the same locality only once in centuries. Consequently, it will not do to sit down quietly at home and wait for one to come, but a person must be up and doing and on the chase.

Rebecca R. Joslin (1929)

10

Observing a Total Eclipse

A total eclipse of the Sun. What is this sight that lures people to travel great distances for a brief view at best and a substantial possibility of no view at all, with no rain check? And how do you get the most out of the experience of a total eclipse?

In these pages, a dozen astronomers, professionals and amateurs, plus some eclipse seekers from earlier times share their experiences. Among them, they have witnessed a hundred total eclipses.

"It's like a religious experience," says Jay Anderson, "the anticipation as the time until totality is counted in days, then hours, then minutes. It's the perfect build-up. Spielberg couldn't do it better. It's an intensely moving event."

Steve Edberg agrees: "It is the intensity of the event. You grab as much as you can. I like action in the heavens and you can't get much better than this."

First Contact

There is a special feeling from the instant when the Moon begins to slide in front of the Sun. In less than a minute, observers with small telescopes see the first tiny "bite" out of the western side of the Sun. For Virginia Roth, "That's when the magic starts."

First contact remains a very special moment for Jay Pasachoff.

As an astronomer, he can appreciate more readily than most all the factors that go into predicting precisely when an eclipse will occur and exactly where on Earth it will be seen. When the call "First Contact" comes right on schedule, he always finds this ingenuity of mankind astounding.

For Alan Fiala, this accuracy is very personal. He is the U.S. Naval Observatory astronomer responsible for predicting the times and paths of eclipses that appear in the *Astronomical Almanac*. "The aspect of anticipation that I feel most keenly is that an eclipse occurs just as I predicted, because so many people rely on my predictions."

Even a century and a half ago, this commencement of a rare event was already exerting a powerful effect on its beholders. French astronomer François Arago observed the 1842 eclipse from Perpignan in southern France amid townspeople and farmers who had been edu-

---

Panel of Eclipse Veterans

Jay Anderson, meteorologist, Prairie Weather Centre, Winnipeg, Canada
John R. Beattie, typesetter, New York City
Richard Berry, editor-in-chief, *Astronomy,* Milwaukee, Wisconsin
Dennis di Cicco, associate editor, *Sky & Telescope,* Cambridge, Massachusetts
Stephen J. Edberg, astronomer, Jet Propulsion Laboratory, Pasadena, California; and past president, Western Amateur Astronomers
Alan D. Fiala, chief, Astronomical Data Division, Nautical Almanac Office, U.S. Naval Observatory, Washington, D.C.
Ruth S. Freitag, senior science specialist, Library of Congress, Washington, D.C.
George Lovi, astronomy author/columnist and planetarium educator, Lakewood, New Jersey
Frank Orrall, professor of physics and astronomy, Institute for Astronomy, University of Hawaii, Honolulu
Jay M. Pasachoff, Field Memorial Professor of Astronomy and director of the Hopkins Observatory, Williams College, Williamstown, Massachusetts
Leif J. Robinson, editor, *Sky & Telescope,* Cambridge, Massachusetts
Virginia and Walter Roth, proprietors, Scientific Expeditions, Inc. (eclipse tour organizers), Venice, Florida
Jack B. Zirker, astronomer, National Solar Observatory, Sunspot, New Mexico

---

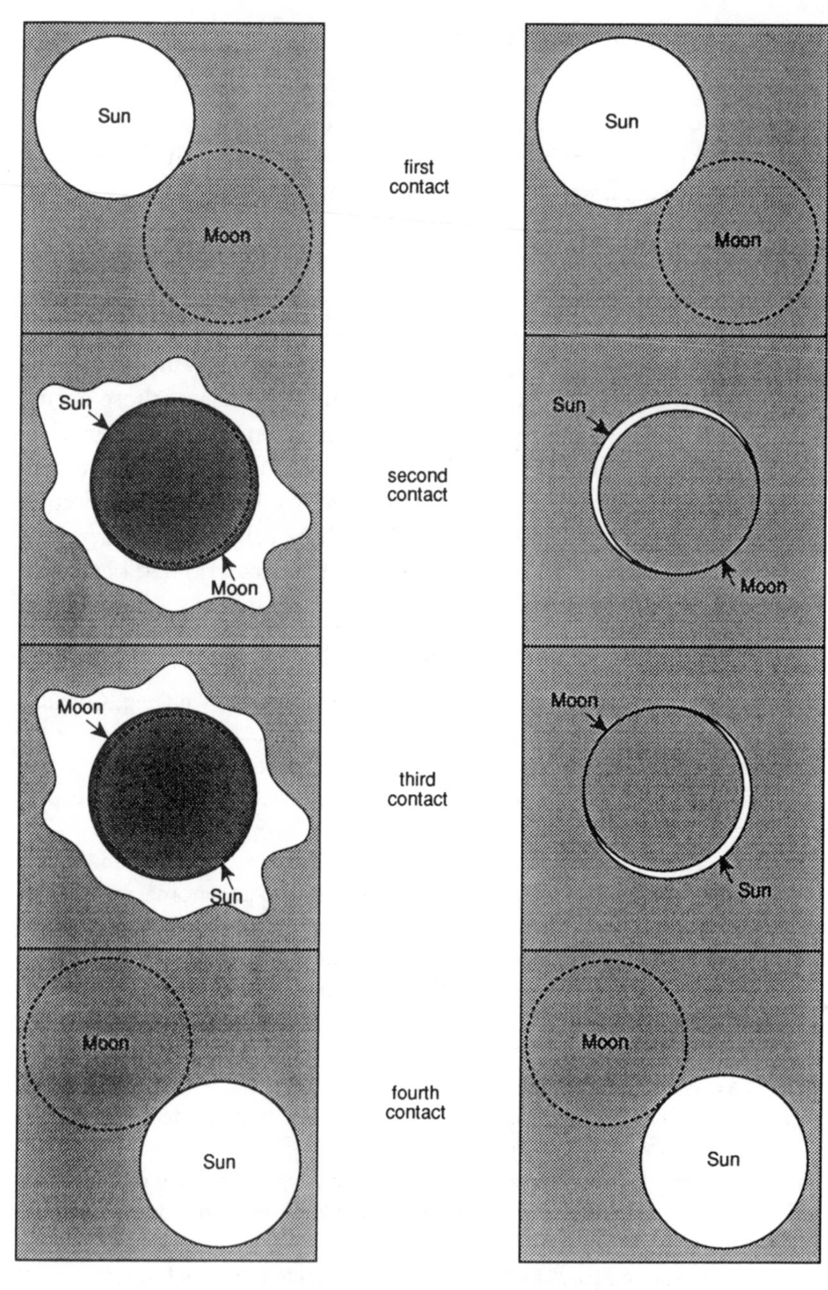

Points of contact in a total solar eclipse *(left)* and an annular eclipse *(right)*. When limb of the Moon is not silhouetted against the Sun it is invisible and is represented by dashes. At second and third contacts in a total eclipse the Sun is hidden behind the Moon, and its limb is also represented by dashes. In a partial eclipse there are only first and fourth contacts.

cated about the eclipse and who were watching the sky intently. "We had scarcely, though provided with powerful telescopes, begun to perceive a slight indentation in the sun's western limb, when an immense shout, the commingling of twenty thousand different voices, proved that we had only anticipated by a few seconds the naked eye observation of twenty thousand astronomers equipped for the occasion, and exulting in this their first trial."[1]

The Crescent Sun

The partial phase of a total eclipse has a power all its own, as the Moon steadily encroaches upon the Sun, covering more and more of its face. A partial eclipse close to total visited a Russian monastery in medieval times, near sunset on May 1, 1185. A chronicler recorded: "The sun became like a crescent of the moon, from the horns of which a glow similar to that of red-hot charcoals was emanating. It was terrifying to men to see this sign of the Lord."[2]

As the partial phase proceeds, a thousand tiny images of the crescent Sun may be visible on the ground beneath trees as the spaces between the leaves create pinhole cameras to focus the crescent image of the waning Sun. Yet, as George B. Airy, astronomer royal of England, learned when he saw his first total eclipse in 1842, "[N]o degree of partial eclipse up to the last moment of the sun's appearance gave the least idea of a total eclipse . . . ."[3]

Crescent images of the partially eclipsed Sun formed by foliage

The Changing Environment

Every eclipse veteran urges newcomers to pause as totality approaches to look around at the landscape and notice the changes in light levels and color. Many people are surprised how little the landscape darkens until the last ten minutes or so before the eclipse becomes total. All the better for the drama, because once the light begins to fade, it sinks noticeably. Usually, everyone becomes quite silent. You can almost feel the tension and rising emotion. A primitive portion of your brain tugs at you to say that something peculiar is going on, that it ought not to be growing dark in the midst of day.

Dennis di Cicco remembers that during the 1976 eclipse in Australia, the birds responded to the fading light by raising a racket and going to roost. In 1973, he witnessed an annular eclipse in Costa Rica that took place shortly after sunrise. The cows grazing around him ignored the partial phases of the eclipse, but as the eclipse reached maximum, the cows ceased grazing, formed a line, and marched back to their barn.

Walter Roth recalls watching the 1973 eclipse from a game preserve in Africa. As the light faded in the minutes just before totality, the birds flocked into the trees, complaining madly. Elephants, which had been grazing peacefully, milled around nervous and confused.

Catching Shadow Bands

by Laurence A. Marschall

Even though shadow bands are only visible for a few fleeting minutes, it is possible to catch them in flight if you prepare in advance. Get a large piece of white cardboard or white-painted plywood to act as a screen—the bands are subtle and can be more easily seen against a clean, white surface. A large white sheet staked to the ground may be more portable and will serve in a pinch, but ripples in the sheet can mask the faint gradations of the shadow bands.

Lay on the screen one or two sticks marked with half-foot intervals (yardsticks will do nicely). Orient the sticks at right angles to one another so that at the first sign of activity you can move one stick to point in the direction that the shadow bands are moving. Then, using the marks on that stick, make a quick estimate of the spacing between the bands (typically 4 to 8 inches). Finally, using a watch, make a quick timing of how long a

"As the light fades, the Sun is a thin crescent, and shrinking quickly," says Richard Berry. "The quantity and quality of the light have begun to change noticeably. The temperature is dropping; the air feels still and strange."

Shadow Bands

As the eclipse nears totality and shortly after it emerges from totality, shadow bands—faint undulations of light rippling across the ground at jogging speed—are sometimes visible. They are one of the most peculiar and least expected phenomena in a total eclipse. Many eclipse veterans have never seen them; some do not want to take time to try because they occur in the last moments before totality as Baily's Beads and other beautiful sights are visible overhead. Other veterans consider shadow bands one of the true highlights of a total eclipse. They resemble the graceful patterns of light that flicker or glide across the bottom of a swimming pool. Shadow bands occur when the crescent of the remaining Sun becomes very narrow so that only a thin shaft of sunlight enters the atmosphere of Earth overhead. There it encounters currents of warmer and cooler air that have slightly different densities. The different densities act as very weak lenses to bend the light passing from one parcel to the next. It is the focusing and

---

bright band takes to go a foot or a yard. Jot down the figures or, better yet, dictate your measurements into a small tape recorder. If you practice this procedure before the eclipse, you will be able to see the shadow bands and then swing your attention back to the sky to catch Baily's Beads and the onset of totality.

The second stick, by the way, is reserved for measuring shadow bands *after* totality, if they are visible.

Later, when the excitement of the eclipse is over, you can take stock of the data. Do the shadow bands seem to move at all? At some eclipses, especially when the air is very still, they just shimmer without going anywhere. If they move, how fast? Typical speeds are about 5 to 10 miles an hour. Do they change directions after eclipse? Usually they do, unless you happen to be standing directly along the centerline of totality.

Shadow bands on an Italian house in 1870

defocusing of light by these ever-present air currents as they are blown about by the wind that causes stars to twinkle. The Sun would twinkle too if it were a starlike dot in the sky. And so it does, near the total phase of a solar eclipse. When the Sun has narrowed from a disk to a sliver, the light from the sliver twinkles in the form of shadow bands rippling across the ground.

In 1842, George B. Airy, the English astronomer royal, saw his first total eclipse of the Sun and recalled shadow bands as one of the highlights: "As the totality approached, a strange fluctuation of light was seen . . . upon the walls and the ground, so striking that in some places children ran after it and tried to catch it with their hands."[4]

The Approach of Totality

During the final one to two minutes before totality, the Sun is about 99 percent covered and the light is fading rapidly. In these brief moments, Baily's Beads, the Diamond Ring, and the corona all appear.

Looming on the western horizon, growing ever larger, is the Sun-cast dark shadow of the Moon. It is coming toward you. Alan Fiala describes its appearance as the granddaddy of thunderstorms, but utterly calm. If you are observing from a hill with a view to the west, the approach of the Moon's shadow can be quite dramatic, even chilling. In the words of astronomer Mabel Loomis Todd a century ago, it is "a tangible darkness advancing almost like a wall, swift as imagination, silent as doom."[5] And this is how astronomer Isabel Martin Lewis in 1924 described the onrush of the Moon's shadow at the onset of totality: "[W]hen the shadow of the moon sweeps over us we are brought into direct contact with a tangible presence from space beyond and we feel the immensity of forces over which we have no control. The effect is awe-inspiring in the extreme."[6]

As the shadow races east toward you, it accelerates. In the last few seconds before totality, you feel as if you are being swallowed by some gigantic whale. In these seconds, as the soft white glow of the corona begins to silhouette the dark disk where the face of the Sun once shone, the ends of the remaining sliver of the Sun fracture into Baily's Beads. Each bead lasts only an instant and flickers out as new ones form. As the length of the Sun's sliver shortens, the two separated groups of beads converge and combine, and suddenly only one

Diamond Ring Effect (June 11, 1983). © *1983 Stephen J. Edberg*

remains—the Diamond Ring. "For one fleeting moment this last bead lingers, like a single jewel set into the arc that is the lunar limb," says John Beattie. And then it is gone.

The eclipse is total.

Corona Emerging/Second Contact

It is difficult to find words that do justice to the corona, the central and most surprising of all features of a total eclipse. The Sun has vanished, blocked by the dark body of the Moon—a black disk surrounded by a white glow; "the eye of God," in the words of Jack Zirker.

In the bitter cold of the high plateau in Bolivia, astronomer Frank Orrall was part of a research team studying the total eclipse of 1966. In reviewing his observing notes, he realized that he had written "The heavens declare the glory of God." "My notebooks normally do not say such things," he mused.

Even the color of the corona is difficult to capture. "I prefer 'pearly,' " says Ruth Freitag, "because it conveys a luminous quality that 'white' lacks." "A silver gossamer glow," suggests Edberg. "Remember," says George Lovi, "the human eye is the only instrument that can see the corona in all its splendor."

During Totality

In the midst of totality, it is particularly impressive to look around at the nearby landscape and the distant horizon. Depending on your position within the eclipse shadow, the size of the shadow, and cloud conditions away from the Sun's position, the light level during totality can vary greatly from eclipse to eclipse and from place to place within an eclipse. On some occasions, totality brings the equivalent of twilight soon after sunset; on others it is dark enough to make reading difficult. Yet it is never as black as night. The color that descends upon everything is hard to specify. Most veteran observers describe it as an eerie bluish gray or slate gray, and then apologize for the inadequacy of the description. The Sun's corona contributes some brightness, but only about as much as a Full Moon. Primarily, it is light reflected from a short distance away, where the Sun is not totally eclipsed, that brightens your location within the shadow of the Moon.

Looking around the horizon in the midst of totality, you have the best impression of standing in the shadow of the Moon. You also have a renewed appreciation for the power of the Sun. There it is, its face completely blocked by the Moon. How big in the sky is that body whose overwhelming brightness creates the day and banishes the stars from view? Stretch out your arm toward the eclipsed Sun as far as you can comfortably reach, just as you did to measure the Moon. Squeeze the darkened Sun between your index finger and your thumb until it just fits between them, as if you were going to pluck it from the sky. It is the size of a pea. Yet that little spot in the sky, darkened now, is usually enough to blanket the Earth in light and warmth and completely dazzle the eye. For just a few minutes its face is hidden. Daytime has become night and the temperature is falling. Do you feel something of what ancient people must have felt as they watched the Sun, on which they depended for warmth and light and life itself, disappear?

Leif Robinson emphasizes that a total eclipse cannot be experienced vicariously. Even the best photography and video tapes are pale reflections of the event. Taking pictures during an eclipse is fine, but be sure you *look* at what is truly a visual spectacle. Do not miss the ambiance of the moment either, he advises. Look at other people to see how they are reacting. Notice the changing colors in the sky. He does not, however, recommend trying to see planets or stars in the sky during a total eclipse. "Time during totality is too precious to spend straining to see stars when you can see those same objects much better in the nighttime sky." Edberg feels differently: "One of my strongest memories from eclipses is seeing stars during the day."

The transformation in the appearance of the Sun from a bright crescent to a dark disk surrounded by ghostly light was recorded by François Arago, the dean of French astronomy a century and a half ago, as he watched the eclipse of July 8, 1842, with a group of astronomers and nearly 20,000 townspeople in southern France.

> [W]hen the sun, being reduced to a narrow filament, began to throw only a faint light on our horizon, a sort of uneasiness took possession of every breast, each person felt an urgent desire to communicate his

emotions to those around him. Then followed a hollow moan resembling that of the distant sea after a storm, which increased as the slender crescent diminished. At last, the crescent disappeared, darkness instantly followed, and this phase of the eclipse was marked by absolute silence . . . The magnificence of the phenomenon had triumphed over the petulance of youth, over the levity affected by some of the spectators as indicative of mental superiority, over the noisy indifference usually professed by soldiers. A profound calm also reigned throughout the air: the birds had ceased to sing.

After a solemn expectation of two minutes, transports of joy, frenzied applauses, spontaneously and unanimously saluted the return of the solar rays. The sadness produced by feelings of an undefinable nature was now succeeded by a lively satisfaction, which no one attempted to moderate or conceal. For the majority of the public the phenomenon had come to a close. The remaining phases of the eclipse had no longer any attentive spectators beyond those devoted to the study of astronomy.[7]

The View from the Edge

by Alan D. Fiala

The advice in this chapter assumes that the observer is near the central line of an eclipse. But observing a total eclipse from the center of its path is not the only possibility.

At any total eclipse, if you move toward either edge of the path of totality, you gain in the duration of Baily's Beads activity. The number of beads you see depends on whether you go toward the southern or northern limit of the eclipse path. The southern edge provides more beads because the terrain of the Moon's southern limb is much rougher. The bead activity at the edges of the path is always caused by the same lunar features, so you can control the amount of bead activity you see by how close you come to the correctly predicted limit.

If you observe from the center line of the eclipse, the beads are confined to the eastern limb of the Moon as totality approaches and the western edge of the Moon after the end of totality. Bead activity at the eastern and western limbs is strongly affected by lunar librations, which change the view significantly from one eclipse to another and determine whether or not at second or third contact there is any Diamond Ring Effect at all.

If you want to study the inner corona or the corona near one pole (where the corona tends to appear more brushlike), once again you gain by

How do you take it all in? It is not possible. There is too much to see—and feel. "Everybody, myself included, tries to do too much," says Edberg. "Save time near the beginning, middle, and end of totality just to stare," urges Anderson. "Make a deliberate effort to store the sights in your mind." Mrs. Todd wrote, a century ago, that when Dr. Peters of Hamilton College was asked what single instrument he would select for observing an eclipse, he replied, "a pillow."[8]

The End of Totality

All too soon, no matter what the duration of totality (and it can never exceed 7 minutes 31 seconds), the Moon's shadow moves on and a bright flash of sunlight appears from the western edge of the Moon.

Rebecca Joslin and her college astronomy teacher traveled from the United States to Spain for an eclipse in 1905, only to be clouded

---

going toward the edge of the path of totality. In this way, you control what portion of the inner corona you observe.

The center line is not the only place from which to enjoy a total eclipse of the Sun.

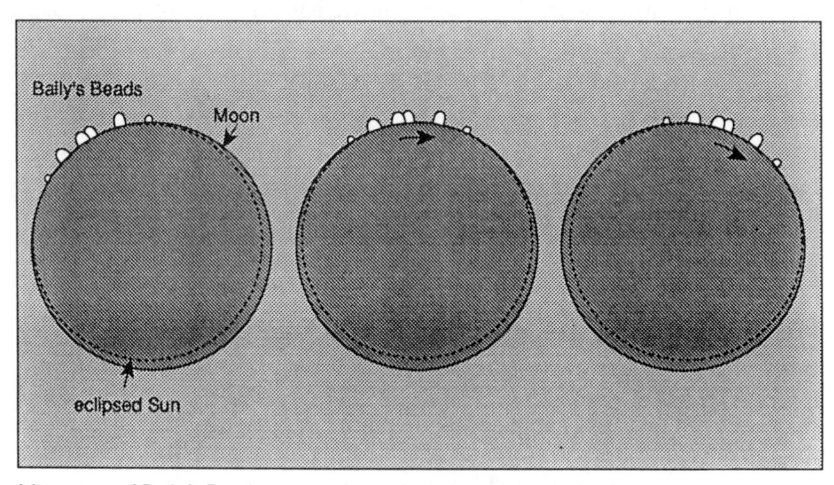

Migration of Baily's Beads as seen from the edge of the path of totality

out. They shifted their attention to nature around them until light pierced the cloud to tell them that the unseen total phase of the eclipse was over.

> But we hardly had time to draw a breath, when suddenly we were enveloped by a palpable presence, inky black, and clammy cold, that held us paralyzed and breathless in its grasp, then shook us loose, and leaped off over the city and above the bay, and with ever and ever increasing swiftness and incredible speed swept over the Mediterranean and disappeared in the eastern horizon.
>
> Shivering from its icy embrace, and seized with a superstitious terror, we gasped, "WHAT was THAT?" . . . The look of consternation on M's face lingered for an instant, and then suddenly changed to one of radiant joy as the triumphant reply rang out, "THAT was the SHADOW of the MOON!"[9]

"I doubt if the effect of witnessing a total eclipse ever quite passes away," wrote Mrs. Todd. "The impression is singularly vivid and quieting for days, and can never be wholly lost. A startling nearness to the gigantic forces of nature and their inconceivable operation seems to have been established. Personalities and towns and cities, and hates and jealousies, and even mundane hopes, grow very small and very far away."[10]

"Beware of post-eclipse depression," warns Edberg. One minute after third contact, with the passage of the Moon's shadow, the disappearance of the shadow bands, and the reappearance of the crescent Sun, you feel exhausted: worn out by totality and the excitement of getting ready for it. Most observers are too tired to watch after third contact, and what follows is anticlimactic anyway. The cure for blue sky blues? "Socialize," says Edberg. "Ask people what they saw. Share war stories."

"The end of totality is a time for celebration," says Anderson. "In the wake of such a powerful shared experience, conversations become animated and casual acquaintances tend to become lifelong friends."

Charles Piazzi Smyth, astronomer royal of Scotland, saw his first total eclipse in 1851, and recognized its ability to overwhelm an observer, distracting him from his carefully planned research. "Although it is not impossible but that some frigid man of metal nerve may be found capable of resisting the temptation," he wrote, "yet certain it is that no man of ordinary feelings and human heart and soul can withstand it."[11]

Confessions of Eclipse Junkies

What happens if you are clouded out, as happens even with good planning about one out of every six times? Unlike a rocket countdown, which can be "T minus 30 minutes and holding" while the clouds clear, an eclipse countdown is inexorable, and it is not always possible to race to another location where there is a break in the clouds.

There were high hopes for the total eclipse of August 19, 1887, but clouds spoiled the view along almost the entire path from Germany through Russia to Japan. Someone posted a public notice in Berlin stating that, on account of the weather, the eclipse had been postponed to another day![12]

Bad luck plagued Rebecca Joslin in her eclipse chasing. She started as a college student by sailing off to Spain in 1905, a journey that meant weeks of travel. The weather was perfect until about 10 minutes before totality, when a single cloud appeared, blotted out the

Clouded out. Glum observers at an eclipse in Lapland (August 9, 1896) see only the twilight glow on the horizon in this painting by Lord Hampton.

The Personalities of Eclipses

by Stephen J. Edberg

Weather, geography, and companionship profoundly affect the experience of an eclipse. But each eclipse has intrinsic differences from all others that continue to lure eclipse veterans. They use these differences in planning their observations.

One factor is the magnitude of the total eclipse, the degree to which the disk of the Moon more than covers the disk of the Sun. When the angular size of the Moon is great, the Moon at mid-totality will mask not only the Sun's photosphere, but also its chromosphere and lowermost corona. Except at the beginning and end of totality (or near the eclipse path limits), the stunning fluorescent pink of the prominences will be hidden from view, unless an absolutely gigantic prominence happens to be present on the Sun's limb.

However, because the relatively bright lower corona is blocked from view, a large-magnitude eclipse is the best time to observe the full extent and detail in the corona. Because the Moon appears larger, its shadow is wider. Thus the total eclipse lasts longer and the darkness is deeper, allowing the full majesty of the corona to shine through as well as any planets and bright stars that happen to be above the horizon.

By contrast, in a smaller-magnitude total eclipse, the Moon's disk is not big enough to mask the lower corona, and prominences may be seen all the way around the disk of the Sun during most of totality. But this eclipse will be comparatively brief and the full extent of the corona may not be evident. A total eclipse with less obscuration often provides a better display of Baily's Beads and the Diamond Ring Effect. Because the Moon's disk is nearly the same apparent size as the Sun's disk, as totality nears, the crescent of the Sun is long and narrow, allowing more Baily's Beads to form and glimmer like jewels on a necklace. When the Moon's apparent disk is large, the length of the Sun's crescent is greatly shortened as it narrows. There may be only one or two Baily's Beads.

As veteran observers plan for upcoming eclipses, they also take into consideration the sunspot cycle. At sunspot maximum, the corona is brighter, rounder, and larger. Prominences also tend to be more numerous. At sunspot minimum, the corona is fainter and extends farther from the equator than the poles. Brush-like coronal features projecting from the poles are more noticeable.

A third factor considered by eclipse followers is the proximity of the shadow path to the Earth's equator, where the rotation of the Earth causes the ground speed of the Moon's shadow to be slowest and thus the length of totality is greater. Eclipses with six to seven minutes of totality are almost always found between the Tropic of Cancer and the Tropic of Capricorn. The rotation of the Earth prolongs not just the period of totality but all aspects of the eclipse, so that, near the equator, the partial phases of the eclipse last longer. The last crescent of the Sun is covered more slowly, so Baily's Beads and the Diamond Ring Effect, while still brief, last a little longer, and the view of the prominences at the limb of the Sun is also prolonged.

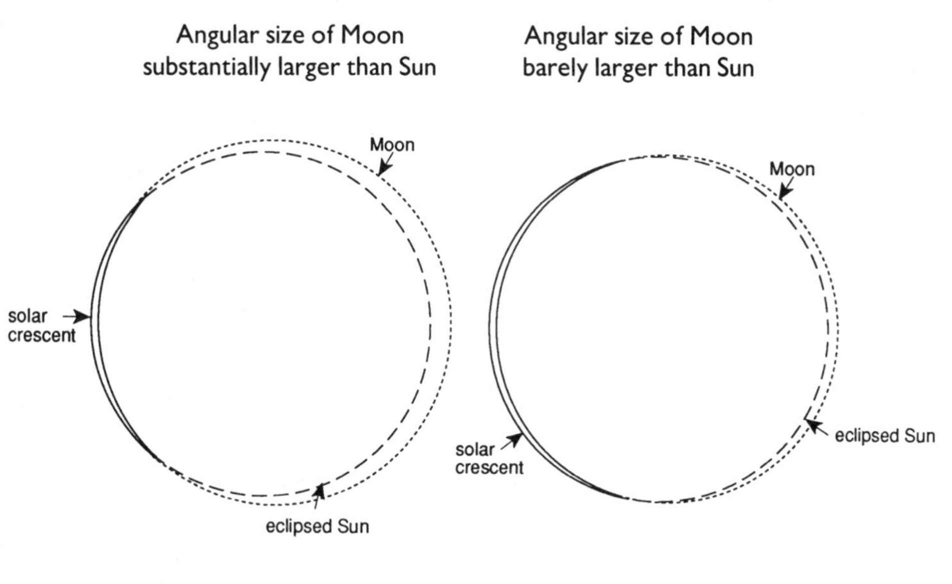

Left: when the Moon's angular size is large, the Sun's crescent is short, reducing the span of Baily's Beads. *Right:* when the Moon just barely covers the Sun, so that the disks appear nearly the same size, the Sun's crescent is much longer and the span of Baily's Beads is greater. (The figure on the left illustrates the solar crescent for the 1991 eclipse less than one minute before totality.)

Sun for the entire 3½ minutes of totality, and then moved on. She tried again in August 1914, sailing first for England, with reservations on a ship that would take her to a viewing site in Norway. While she was enroute to England, World War I broke out, her England-to-Norway cruise was cancelled, and she watched a partial eclipse from England under absolutely clear skies.

It was not until January 24, 1925, 20 years after her first attempt, that Joslin finally saw a total eclipse. This time the eclipse came almost to her. She had only to cross from Massachusetts into Connecticut. She had previously been wiped out by one cloud and one world war. Nature did not make it easy on her this time either. The temperature at her site was –2 °F. One brief eclipse out of three. She gathered her travels and experiences into a book called *Chasing Eclipses*. Because of her keen eye for people, places, history, and irony, she salvaged quite a bit from misfortune.

All the veterans agree: Plan your eclipse trip so that you see things, go places, and meet people that will shine in your memory even if the corona does not.

Crucial to the enjoyment of a solar eclipse is eye safety. You would not stare directly at the Sun during a normal day, so you should not stare at the Sun when it is partially eclipsed without proper eye protection (described in the next chapter). However, when the Sun is totally eclipsed—no portion of its disk is showing—it is perfectly safe to look at the Sun without any eye protection. Pasachoff is irritated by governments and news media in foreign lands and even in this country that mislead the public by implying that the Sun gives off "special rays" at the time of an eclipse. Instead of teaching simple safety procedures, they frighten, thereby depriving people of a rare and magnificent sight.

Fiala recalls the 1980 eclipse in India, where pregnant women were instructed to remain indoors. He, Lovi, and others remember the 1983 eclipse in Indonesia. Before the eclipse the streets were teeming with people. On eclipse morning, however, there was scarcely anyone outdoors. Fire sirens wailed like an air raid warning for people to take cover. School children, on government orders, were kept indoors, forbidden to see the eclipse, except perhaps on television. A stunning natural event that comes to you, if you are lucky, once in five lifetimes, was denied to them.

On Java for the 1983 eclipse, Anderson recalled soldiers patrolling the streets to discourage unauthorized observers, although the citizens were so frightened by government warnings that almost all stayed indoors. As the Sun was gradually disappearing, the soldiers near him began to be caught up in the excitement. Anderson offered them a view through his telescope, but the warnings they had received overtook them and they refused. However, several observers gave the soldiers a few minutes of careful explanation, and they nervously took a peek. "After their first look all apprehension disappeared, and they participated in the event as fully as we did. Ten minutes before totality, the officer in charge was interrupted by a call on his walkie-talkie from his superior at headquarters:

'How are things going out there?'

'O.K. There are no problems.'

'It's time to come in now. Collect your men and bring them back to the barracks. We are supposed to have all troops back before the eclipse begins.'

'There are no problems here; everyone is safe; the tourists are all working on their equipment. Are you sure you want us in: I see no problems.'

'Orders are to bring everyone in, and leave the tourists to themselves. Bring the men in.'

'I'm sorry, I can't hear you. I'm having trouble with the radio. What did you say?'

At this point he turned off the radio and put it back in his pocket. The entire troop stayed to watch the eclipse with us."

Even more memorable to Anderson were two dozen Indonesian students from a local college. They were studying English and had attached themselves to Anderson's group to practice their speaking skills. As totality neared, the students became extremely nervous. To reassure them, Anderson and his wife held hands with them during totality—the most magical moment in all the eclipses he has seen.

Not only in underdeveloped countries do superstition and misinformation reign supreme and deprive people of a rare gift of nature. The eclipse of 1970 was the first and most impressive of many Lovi has seen. It was visible in the eastern United States and he was watching from Virginia Beach. The news media had trumpeted the event, but placed great emphasis on the dangers. Around him were thousands of people, most of them casual observers, watching as the crescent Sun thinned and vanished. Instantly a cry went up, spreading from group to group: "Look away! Look away!" Lovi knows of other curious people who traveled substantial distances to reach Virginia Beach and then, fearful, stayed in their motel rooms and watched the event on television.

Di Cicco was in a cornfield in North Carolina for the 1970 eclipse and recalls that in the midst of totality, a car came up the road with its lights on and drove right by without stopping, its passengers oblivious or impervious to the wonder above them. Di Cicco also remembers Australia in 1976, where government warnings that people should stay in their houses to avoid the dangers of the eclipse were so intense that he and other observers feared that local citizens or the police might stop them from watching the eclipse out of concern that they were hurting themselves.

Edberg was in Winnipeg, Canada, for the eclipse of 1979. Students had to have notes from their parents to be allowed outside during the eclipse. Those without parental permission were confined to their classrooms, where the shades were drawn and the students watched the eclipse on television. Pasachoff recalls that "One school in Winnipeg even asked for permission to ignore fire alarms if any sounded during the eclipse, lest the students rush outside and be blinded." On a more rational note, Anderson remembers that many parents kept their children home from school on eclipse day and took the day off themselves so that they could witness the eclipse as a family.

The astronomy popularizer Camille Flammarion, quoting Arago, tells of the eclipse of May 22, 1724, visible from Paris. Supposedly, a marquis and some aristocratic lady friends were invited to observe the eclipse at the Paris Observatory. The ladies, however, fussed so long with their gowns and hair styles that the party arrived late, a few minutes after totality had ended. "Never mind, ladies," said the marquis, "we can go in just the same. M. Cassini [the observatory director] is a great friend of mine and he will be delighted to repeat the eclipse for you."[13]

Stages of a Total Eclipse

First Contact The Moon begins to cover the western limb of the Sun.

The Crescent Sun Over a period of about an hour, the Moon obscures more and more of the Sun, as if eating away at a cookie. The Sun appears as a narrower and narrower crescent.

Light and Color Changes About 15 minutes before totality, when 80 percent of the Sun is covered, the light level begins to fall noticeably, and then it falls with increasing rapidity and the landscape takes on a metallic gray-blue hue.

Gathering Darkness on the Western Horizon About 15 minutes before totality, the shadow cast by the Moon becomes visible on the western horizon as if it were a giant but silent thunderstorm.

Animal, Plant, and Human Behavior As the level of sunlight falls, animals may become anxious or behave as if nightfall has come. Some plants close up. Notice how the people around you are affected.

Temperature As sunlight fades, the temperature may drop perceptibly.

Shadow Bands A few minutes before totality, ripples of light may flow across the ground and walls as the Earth's turbulent atmosphere refracts the last rays of sunlight.

Corona Up to one minute before totality, the corona begins to emerge.

Baily's Beads About 10 seconds before totality, the Moon has covered the entire face of the Sun except for a few rays of sunlight passing through deep valleys at the Moon's limb, creating the effect of jewels on a necklace.

Shadow Approaching While all this is happening, the Moon's dark shadow in the west has been growing. Now it rushes forward and envelops you.

Diamond Ring Effect The Moon covers all but one of Baily's Beads and that bead vanishes, often as if sucked into an abyss.

Second Contact Totality begins. The Sun's photosphere is completely covered by the Moon.

Prominences and the Chromosphere For a few seconds after totality begins, the Moon has not yet covered the lower atmosphere of the Sun, and a thin strip of the vibrant red chromosphere is visible at the Sun's eastern limb. Stretching above the chromosphere and into the corona are vivid red prominences.

Corona Extent and Shape The corona and prominences vary with each eclipse. How far (in solar diameters) does the corona extend? Is it round or is it broader at the Sun's equator? Does it have the appearance of short bristles at the poles? Look for loops, arcs, and plumes that trace solar magnetic fields.

Landscape Darkness and Horizon Color Each eclipse, depending mostly on the Moon's angular size, creates its own level of darkness. Beyond the Moon's shadow, at the far horizon all around you, the Sun is shining and the sky has twilight orange and yellow colors.

Temperature Is it cooler still? A temperature drop of about 4 °F (2 °C) is typical.

Animal, Plant, and Human Reactions What animal noises can you hear? How do you feel?

Planets and Stars Visible Venus and Mercury are often visible near the eclipsed Sun, and other bright planets and stars may also be visible, depending on their positions and the Sun's altitude above the horizon.

End of Totality Approaching The western edge of the corona begins to brighten and the reddish prominences and chromosphere appear.

Third Contact One bright dot of the Sun's photosphere returns to view. Totality is over. The stages of the eclipse repeat themselves in the opposite order.

Diamond Ring Effect

Shadow Rushes Eastward

Baily's Beads

Corona Fades

Shadow Bands

Crescent Sun

Recovery of Nature

Partial Phase

Fourth Contact The Moon no longer covers any part of the Sun. The eclipse is over.

Shadow and sun—so too our lives are made—
Here learn how great the sun, how small the shade!
Richard Le Gallienne (1920?)

11 Observing Safely

WARNING

Permanent eye damage can result from looking at the disk of the Sun directly, or through a camera viewfinder, or with binoculars or a telescope even when only a thin crescent of the Sun or Baily's Beads remain. The 1 percent of the Sun's surface still visible is about 10,000 times brighter than the Full Moon. Staring at the Sun under such circumstances is like using a magnifying glass to focus sunlight onto tinder. The retina is delicate and irreplaceable. There is nothing a retinal surgeon will be able to do to help you. Never look at the Sun outside of the total phase of an eclipse unless you have adequate protection.

Of course, once the Sun is entirely eclipsed, its bright surface is hidden from view and it is safe to look directly at the totally eclipsed Sun without any filters. In fact, it is one of the greatest sights in nature.

There are five basic ways to observe the partial phases of a solar eclipse without damage to your eyes.

The Pinhole Projection Method

One safe way of enjoying the Sun during a partial eclipse—or anytime—is with a "pinhole camera," which allows you to view a *projected* image of the Sun. There are fancy pinhole cameras you can make out of cardboard boxes, but a perfectly adequate (and portable) version can be made out of two thin but stiff pieces of white cardboard. Punch a small clean pinhole in one piece of cardboard and let the sunlight fall through that hole onto the second piece of cardboard which serves as a screen, held below it. An inverted image of the Sun is formed. To make the image larger, move the screen farther from the

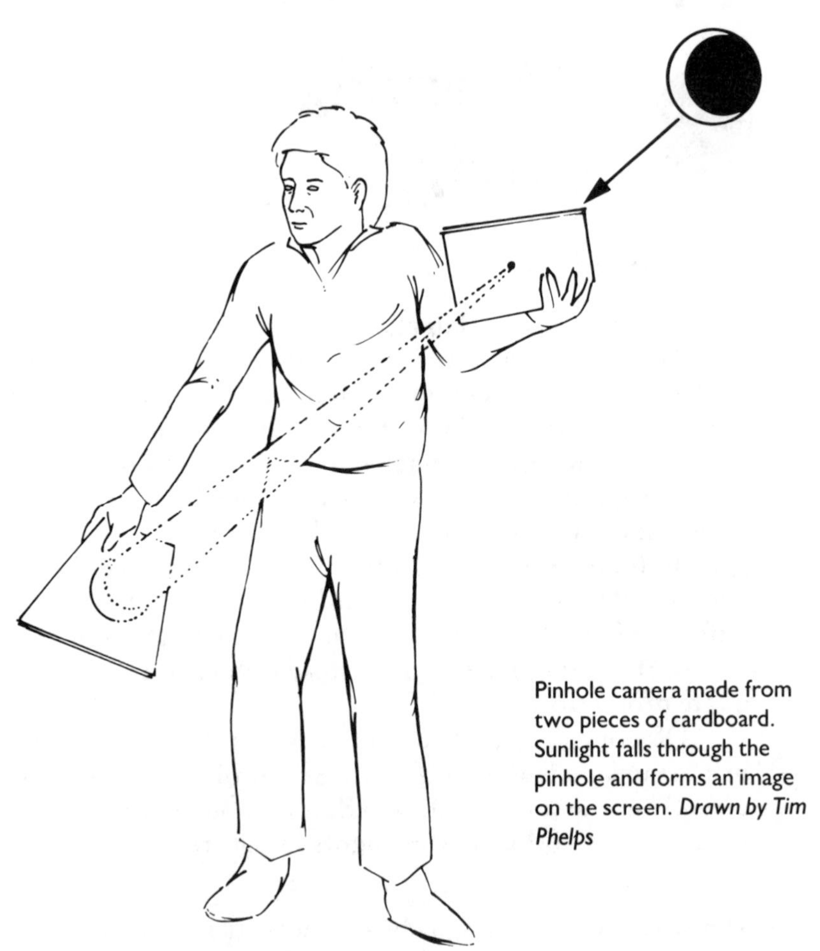

Pinhole camera made from two pieces of cardboard. Sunlight falls through the pinhole and forms an image on the screen. *Drawn by Tim Phelps*

pinhole. To make the image brighter, move the screen closer to the pinhole. Do not make the pinhole wide or you will have only a shaft of sunlight rather than an image of the crescent Sun. Done correctly, a pinhole projection system can even show large sunspots. Remember, this instrument is used with your back to the Sun. The sunlight passes over your shoulder, through the pinhole, and forms an image on the cardboard screen beneath it. Do **not** look through the pinhole at the Sun.

Mylar Sun Filters

A second technique for viewing the Sun safely is by looking directly at the Sun through an aluminized Mylar filter. Advertisements for such filters may be found in popular astronomy magazines. Beware, though, that Mylar, a plastic, comes in various thicknesses and with various coatings. You need a metal coating to save your eyesight and you need to examine the Mylar for small holes that could allow unfiltered sunlight to reach your eyes and damage them. A good solar filter will allow you to look comfortably at the filament of a high-intensity lamp.

When using any filter, however, do not stare for long periods at the Sun. Look through the filter briefly and then look away. In this way, a tiny hole that you miss is not likely to cause you any harm. You know from your ignorant childhood days that it is possible to glance at the Sun and immediately look away without damaging your eyes. Just remember that your eyes can be damaged without your feeling any pain.

Welders' Goggles

Welders' goggles or the filters for welders' goggles with a rating of 14 or higher are safe to use for looking directly at the Sun. They are also relatively inexpensive.

Camera and Telescope Filters

Many telescope and camera companies provide metal-coated filters that are safe for viewing the Sun. They are considerably more

Eye Damage from a Solar Eclipse

by Lucian V. Del Priore, M.D.

The dangers of direct eclipse viewing have not always been appreciated, despite Socrates' early warning that an eclipse should only be viewed indirectly through its reflection on the surface of water. A partial eclipse in 1962 produced 52 cases of eye damage in Hawaii, and a total eclipse along the eastern seaboard of the United States produced 145 cases in 1970. As many as half of those affected never fully recovered their eyesight.

There is nothing mysterious about the optical hazards of eclipse viewing. No evil spirits are released from the Sun during a solar eclipse, and there is no scientific reason for running indoors to avoid "the harmful humors of the Sun." Eye damage from eclipse viewing is simply one form of light-induced ocular damage, and similar damage can be produced by viewing any bright light under the right (or should I say the wrong!) conditions.

Light enters the eye through the cornea and is focused on the retina by the optical system in the front of the eye. Any light that is not absorbed by the retina is absorbed by a black layer of tissue under the retina called the retinal pigment epithelium. The retina is the human body's video camera; it contains nerve cells that detect light and send the electrical signal for vision to the brain. Without it, we cannot see. Most of the retina is

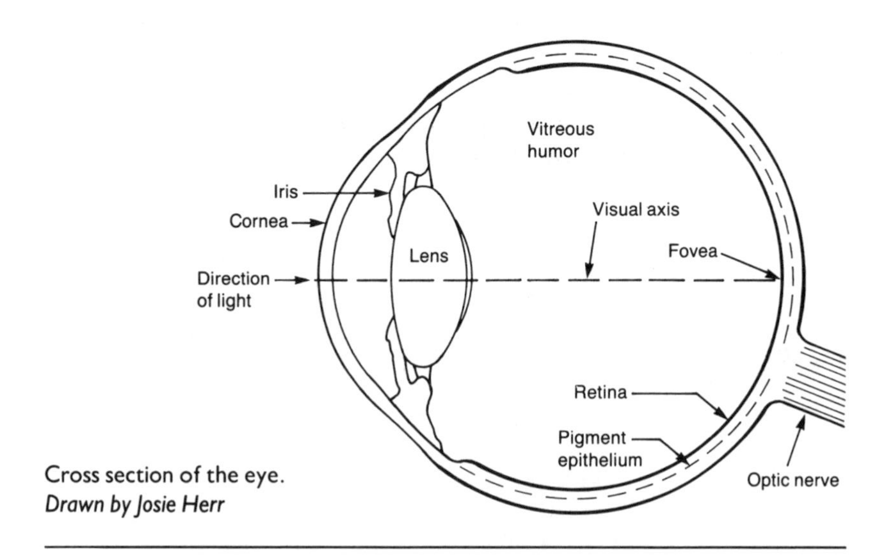

Cross section of the eye.
Drawn by Josie Herr

devoted to giving us low-resolution side vision. Fine detail reading vision is contained in a small area in the center of the retina called the fovea. People who damage their fovea are unable to read, sew, or drive, even though this small area measures only 1/100th of an inch across, and is less than 1/10,000th of the entire retinal area! Unfortunately, this is the precise area that is damaged if we stare directly at the surface of the Sun. Damage has been reported with less than one minute of viewing. The image of the Sun projected onto the retina is about 1/150th of an inch in size, and this is large enough to seriously damage most of the fovea.

Why is sunlight damaging to a structure designed to detect light? Bright sunlight focused on the retina is capable of producing a thermal burn, mainly from the absorption of infrared and visible radiation. The absorption of this light raises the temperature and literally fries the delicate ocular tissue. There is no mystery here, as every schoolchild knows that sunlight focused through a magnifying glass can cause a piece of paper to burst into flames. Yet Sun viewing seldom produces a thermal burn; all but the most intoxicated viewer would surely turn away before it occurs. Instead, most cases of eclipse blindness are related to photochemically induced retinal damage, which occurs at modest light levels that produce no burn and no pain. Two types of light lesions are recognized clinically and experimentally, and both are probably responsible for the damage observed after eclipse viewing. Blue light (400–500 nanometers) damages the retinal pigment epithelium, and leads to secondary changes in the retina, while near ultraviolet light (340–400 nanometers) is absorbed by and directly damages the light-sensitive cells in the outer retina.

Viewing a partial eclipse recklessly is not the only way to produce light-induced retinal damage. Sungazing is a well-known cause of retinal damage, even in the absence of an eclipse. Numerous cases of blindness have been reported in sunbathers, in military personnel on anti-aircraft duty, and in religious followers who sungaze during rituals and pilgrimages. The Sun is not even required to produce light damage; other types of bright lights, including lasers and welders' arcs, will have the same effect. The common thread here is clear: direct viewing of bright lights can damage the retina, regardless of the source. A solar eclipse merely increases the number of potential victims, and brings the problem to public attention.

expensive than common Mylar, but observers generally like them better because they are available in various colors, such as a chromium filter through which the Sun looks orange. Through aluminized Mylar, the Sun is blue-gray. As with the Mylar, you can look directly at the Sun through these filters.

Caution: Do not confuse these filters, which are designed to fit over the lens of a camera or the aperture of a telescope, with a so-called solar *eyepiece* for a telescope. Solar eyepieces are sometimes sold with small amateur telescopes. **They are not safe** because of their tendency to absorb heat and crack, allowing the sunlight concentrated by the telescope's full aperture to enter your eye.

Fully Exposed and Developed Black-and-White Film

You can make your own filter from black-and-white film, but *only* true black-and-white film (such as Kodak Tri-X or Pan-X). Open a roll of black-and-white film and expose it to the Sun for a minute. Have it developed to provide you with negatives. Use the negatives (black in color) for a filter. It is best to use two layers. With this filter, you can look directly at the Sun with safety.

Remember, however, that if you are planning to use black-and-white film as a solar filter, you need to prepare it at least several days in advance.

Caution: Do not use color film or chromogenic black-and-white film (which is actually a color film). Developed color film, no matter how dark, contains only colored dyes, which do not protect your vision. It is the metallic silver that remains in black-and-white film after development that makes it a safe solar filter.

Eye Suicide

Standard or Polaroid sunglasses are not solar filters. They may afford some eye relief if you are outside on a bright day, but you would never think of using them to try to stare at the Sun. So you cannot use sunglasses, even crossed polaroids, to stare at the Sun during the partial phases of an eclipse. They provide little or no eye protection for this purpose.

Observing with Binoculars

Binoculars are astronomer George Lovi's favorite instrument for observing total eclipses. Any size will do. He uses 7 × 50 (magnification of 7 times with 50-millimeter [2-inch] objective lenses). "Even the best photographs do not do justice to the detail and color of the Sun in eclipse, and especially the very fine structure of the corona, with its exceedingly delicate contrasts that no film can capture the way the eye can." The people who did the best job of capturing the true appearance of the eclipsed Sun, he feels, were the nineteenth-century artists who photographed totality with their eyes and minds and developed their memories on canvas.

For people who plan to use binoculars on an eclipse, Lovi cautions common sense. Totality can and should be observed without a filter, whether with the eyes alone or with binoculars or telescopes. But the partial phases of the eclipse, right up through the Diamond Ring Effect, must be observed with filters over the objective lenses of the binoculars. Only when the diamond ring has faded is it safe to remove the filter. And it is crucial to return to filtered viewing as totality is ending and the western edge of the Moon's silhouette begins to brighten. After all, binoculars are really two small telescopes mounted side by side. If observing the Sun outside of eclipse totality without a filter is quickly damaging to unaided eyes, it is far quicker and even more damaging to look at even a sliver of the uneclipsed Sun with binoculars that lack a filter.

Observing with a Telescope

Some observers, such as Ruth Freitag, prefer to watch the progress of the eclipse through a small portable telescope, which offers stability and is much less tiring to use for extended periods than binoculars. A telescope also provides more detail at higher powers, if this is desired. The solar filter is removed at totality, and by switching to the finder scope more of the corona is visible.

A Final Thought

Just remember, says Lovi, "Don't try to do too much. Look at the eclipse visually. Don't be so busy operating a camera that you don't see the eclipse. And don't set off for the eclipse burdened down by baggage and equipment so that you are tired and stressed and too nervous to enjoy the event."

Astronomer Isabel Martin Lewis also warned of the dangers of too many things to do: "A noted astronomer who had been on a number of eclipse expeditions once remarked that he had never SEEN a total solar eclipse."[1]

Each eclipse has at least one phenomenon that
makes it special.

Stephen J. Edberg (1990)

12

Eclipse Photography

It is safe, and thrilling, to look directly at the Sun's corona during a total eclipse. But *only* when the Sun's photosphere is completely covered by the Moon is it safe to view the eclipse without a special solar filter. Looking at the Sun during the partial phases of an eclipse or at any time during an annular eclipse can, without the sensation of pain, cause irreversible damage to your eyes. During partial phases, a filter that blocks out the dangerous ultraviolet and infrared radiation must be used.

Preparations

In retail marketing, the three most important words are location, location, location. The same applies to solar eclipse expeditions.

First: *location*. Choose a site that offers the best possible weather along the eclipse path. If possible, be prepared to move quickly to an alternate site on the morning of the eclipse if the weather poses a problem.

Second: *location*. Inspect the site ahead of time or obtain reliable information from topographic maps or people living in the area about features that may obstruct your view.

Third: *location*. Select a viewing position where it would be diffi-

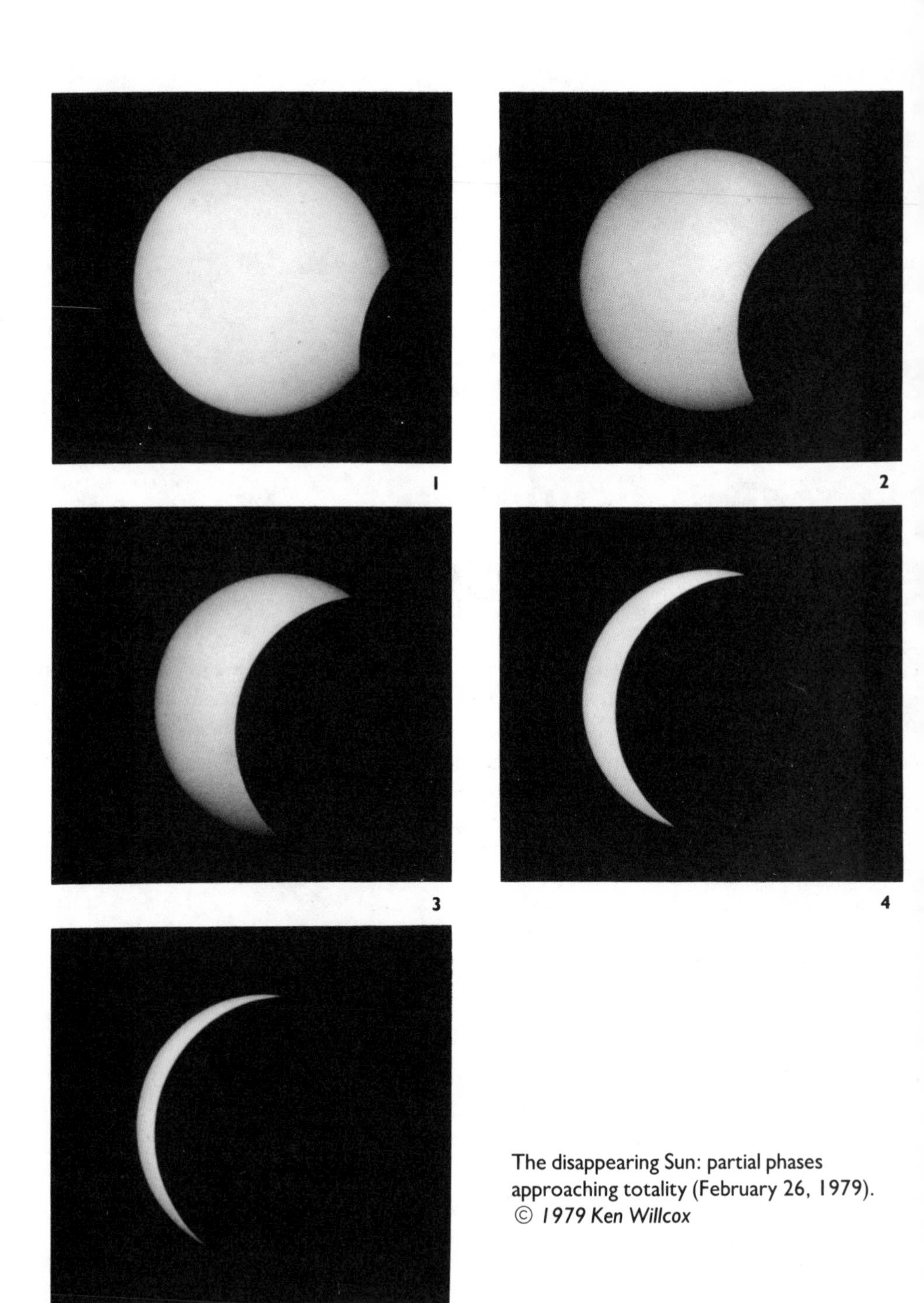

The disappearing Sun: partial phases approaching totality (February 26, 1979). © *1979 Ken Willcox*

cult for someone to move into your line of sight to the Sun during the eclipse. Remember that on the day of the eclipse, there could be several people sharing the same observing site. No matter how well-intentioned, human beings have a habit of moving at just the wrong time.

Because the Sun will be completely covered by the Moon for only a few minutes at most, you should plan and rehearse the photographs you intend to take and know how much time each will require. A written checklist will help you remember what exposure to use and when to remove and replace the solar filter. Do not schedule yourself too tightly. Allow enough time between photographs for any vibrations caused by advancing the film to subside before taking another picture. Remember, substantial darkness accompanies totality and makes it hard to see what you are doing. For that reason, you should have a small flashlight to aid in checking camera settings.

A still greater "hazard," however, is the majesty of the event, which makes even experienced eclipse chasers lose track of what they intended to do. Consider an account from 1842: "[T]he Captain of a French ship had beforehand arranged in the most careful way the observations to be made: but when the darkness came on, discipline of every kind failed, every person's attention being irresistibly attracted to the striking appearance of the moment, and some of the most critical observations were thus lost."[1]

As with most astronomical events, you get only one chance, so take every precaution to be sure you are ready when the time comes. Above all, leave time to absorb the eclipse with all your senses. Do not let your camera be the only eye to see the eclipse. Do not allow all your memories to be on film. A photograph or video recording of a solar eclipse is like viewing a postcard of the Grand Canyon compared to the experience of walking up to the edge and looking down into it for the first time.

As you plan your photography, you first have to decide what kinds of eclipse pictures you want and hence what kind of equipment you will need. Memorable pictures can be taken with a still camera or video camera, with or without a telescope. Several chapters can be written on eclipse photography without exhausting the subject. Here we have space to share only some of the more important ideas.

Film and Solar Filters

Your pictures cannot be any better than your film. There are two basic types of photographic film: print film (negatives) and slide film (transparencies). Unless you have two cameras and four arms, you will be hard pressed to take both prints and slides of a total solar eclipse. An advantage of print film is that it offers a wider latitude of exposures. With print film, good results are less dependent on shutter speed or f-stop. The best slide film is limited to about five f-stops from total black to pure white.[2] Slide film has the advantage of requiring just one processing step, beneficial for retaining accurate color. Slide film also yields a photograph that you can project on a screen for large audiences.

Since the films available for astrophotography are changing at a very rapid pace, it is difficult to rank one over another. Technical bulletins on films applicable to astrophotography are available from film manufacturers. Some of the color print films preferred by amateur astronomers for solar eclipse photography include Kodacolor 400, Fujicolor 400, and Kodak 5072 (sometimes called Vericolor print film, which allows you to make color slides directly from a color negative). Color slide films commonly used for solar eclipses include Ektachrome 400 and Fujichrome 400, both of which can also be used to make beautiful color prints (Cibachromes). If your camera allows for f-ratios less than 5, the use of finer grain films permits you to enlarge

Diamond Ring Effect at eclipse in India (February 16, 1980). © *1980 Jay M. Pasachoff*

your photographs with less loss of detail. Some black-and-white films that are commonly used for solar photography include Kodak TMax 400 (ASA 400), Kodak Technical Pan 2415 (ASA 25–250), Kodak Plus-X (ASA 125), Ilford HP5 (ASA 400), and Agfapan 400 Professional.[3]

Film speed (ASA) should be fast enough to allow you to use a shutter speed of 1/125 second or faster to minimize image drift due to the rotation of the Earth or vibrations caused by subtle winds. This is particularly true if you are using a telephoto lens or telescope to photograph the eclipse.

Solar filters *must* be used to view or photograph the partial phases of the eclipse. No filter is needed during totality. Solar filters vary in material and in the wavelength of light (color) they transmit. In general, aluminized Mylar filters show a blue-gray Sun, while metal-coated glass filters transmit a more realistic orange Sun. Sources for solar filters can be found in advertisements in popular astronomy magazines, such as *Sky & Telescope* and *Astronomy*.

It is critical that you determine well in advance of the eclipse the proper shutter speed and f-ratio for your particular solar filter. Using the camera/lens system, solar filter, and film you intend to use the day of the eclipse, take several photographs of the Sun on different days with different cloud and sky conditions to determine the best exposure through the solar filter. With your solar filter in place, bracket your test shots several f-stops on both sides of what you think might be the proper exposure (for example, 1/30 to 1/500 second), and after viewing the developed photographs choose the settings that give the desired result.

Solar filters for telescopes are made of the same materials used for visual filters, except with a great deal more care taken to assure parallel faces on the filter. There are two types of solar filters: full aperture (including off-axis) and eyepiece filters. Eyepiece filters (such as smoked glass, neutral density, etc.) should be avoided. The tremendous heat generated at the eyepiece can easily shatter the filter. The only safe solar filters for cameras are full-aperture and off-axis filters, both of which fit over the objective of the telescope or camera lens.

Rapid changes in air temperature during a total eclipse as well as slight differences in the parallel alignment of the solar filter and telescope optics can cause changes in focus when the solar filter is removed. Therefore, it is important to check the image focus after

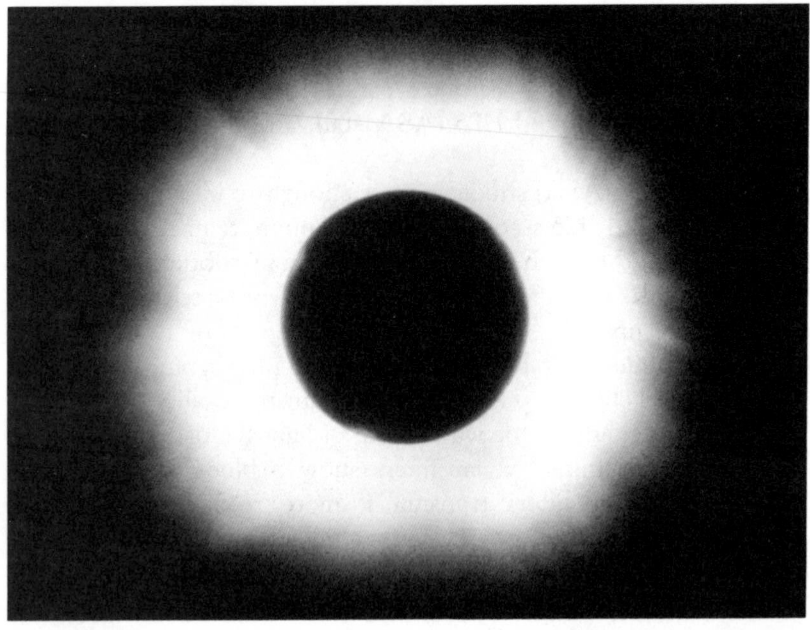

Sun in total eclipse, with corona typical for sunspot maximum (February 26, 1979).

removing the solar filter from the telescope, but not until the Sun is completely covered by the Moon.

If your telescope has a finder scope, be sure to place a solar filter over its objective to keep from burning the cross-hairs. A good idea is to place an inexpensive piece of aluminized Mylar (such as Solar-Skreen) over the objective, or keep the lens cover on the finder scope during the partial phases.

Eclipse Photography without a Telescope

To take *close-up* photographs of the Sun requires at least a 400 mm telephoto lens mounted on a tripod and loaded with fast film (400 ASA or faster). However, if all you have is a normal 50mm lens, you can still obtain spectacular photographs during a solar eclipse—and, because of the wider field of view, you might even discover a new comet! To detect a dim object, such as a comet, which may be several solar diameters from the Sun, take photographs of the area around the Sun at various exposures during totality.

Multiple exposure of the total eclipse of the Sun over Minneapolis (June 30, 1954). Star Tribune, *Minneapolis–St. Paul*

Some of the most dramatic photographs of solar eclipses are multiple exposures that show the totally eclipsed Sun in the center bracketed by a sequence of partial phases as the Sun approaches and emerges from totality. To take a photograph of this remarkable progression of celestial motion, your camera must be capable of taking multiple exposures without advancing the film. Most 35mm reflex cameras on the market today are automatic and will *not* allow this type of photography. Check with the staff of your local camera shop and explain what you want to do; they should be able to show you cameras that will meet your needs. In addition, for multiple exposures you will need a sturdy tripod that will not move when you recock the film shutter between exposures and a cable release to eliminate motion during the exposures.

To record a multiple exposure of the entire eclipse, you need to know the field of view of your camera lens. The Sun moves about 15 degrees per hour, or its own diameter every 2 minutes. Therefore, using a 50mm lens, which has an angle of view of approximately 46 degrees, you can expect the Sun to traverse the entire field in about three hours. Taking an exposure every 5 minutes will provide adequate separation between images on the final photograph.

If you want the image during totality to be in the center of the multiple exposure, you must calculate the times from that point in order to have all of your images equally spaced on both sides of the totally eclipsed Sun. For example, if mid-totality is at 10:32 A.M., then you would add or subtract increments of 5 minutes to that figure to get the times to make your exposures. You may decide to allow more

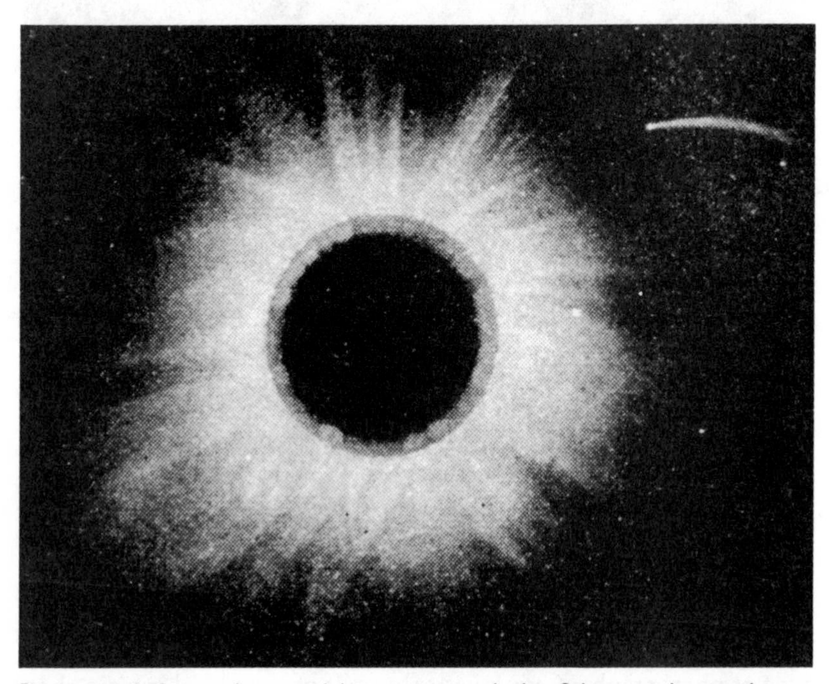

Painting made from a photograph by astronomer Arthur Schuster, who was the first to capture an unknown comet in an eclipse photograph (May 17, 1882)

of an interval just before and after totality to avoid trampling on the corona.

The altitude and azimuth of the Sun or Moon for a specific eclipse are easily obtained using astronomy computer programs available for your home computer.[4] If you do not have a computer, there are several written sources that offer eclipse information several months in advance, such as the U.S. Naval Observatory eclipse circular and *Astronomy* and *Sky & Telescope* magazines. Detailed maps of the eclipse path and information useful for photographing the eclipse are published in these sources.

There are other phenomena associated with an eclipse that can make interesting pictures, such as shadow bands and the crescent images of the Sun produced by tree foliage. During the partial phases of a solar eclipse, the leaves of a tree often provide hundreds of narrow pathways for sunlight to pass. In so doing, they act as "pinhole cameras" to produce crescent images of the partially eclipsed Sun on the

---

Photographing Shadow Bands

by Laurence A. Marschall

Visual observations of shadow bands are interesting and fun, but the most useful observations from a scientific standpoint are those using photography or videotape. The blotchy patterns of the bands can be predicted from mathematical models of how sunlight travels through the Earth's atmosphere. Yet few of the many attempts to catch shadow bands on film or tape have been successful. Perhaps a dozen fuzzy photographs and one flickering film are all that are available. The bands move so swiftly and have such low contrast that it is hard to get a good picture.

To photograph the bands, a high shutter speed (1/250 second or shorter) and very fast film (400 ASA or higher) should be used. Focus the camera on a white cardboard or plywood screen laid out on the ground (see "Capturing Shadow Bands" in chapter 10), and take as many shots as possible, using a range of lens openings to insure a proper exposure. Leave a yardstick on the screen to establish a scale.

For videotaping, many hand-held camcorders have a fast shutter option (1/1000 to 1/4000 second) that should be used to stop the motion of the shadow bands. If you get a distinct record of the shadow bands and their motion, it is not just something to admire; it is a remarkable scientific resource. Shadow band researchers (there are perhaps a half dozen worldwide) will beat a path to your door.

---

ground or the side of a building. Large sunspot groups can sometimes be discerned in this projected image of the Sun. This profusion of crescents on the ground, on a building, or on a person's face can make a nice photographic memento.

"But remember," cautions astrophotographer Dennis di Cicco, "the biggest mistake observers make is that they get so wrapped up in taking pictures that they miss the event." Di Cicco gets his pictures, but always tries to automate his equipment so that he can spend his time just watching the eclipse. He never ceases to be amazed by the beauty. He remembers his first eclipse in 1970 in North Carolina. He was so stunned by the sight that he took a step backward and fell into a ditch full of water.

Calculating Camera Settings

Several factors determine the length of time the camera shutter is open to get the proper image on your final photograph. The table accompanying this chapter provides standard settings for various photographs. For situations not listed, use the following formula to determine exposure.

$$E = f^2/(A \times B)$$

E = exposure (seconds)
f = focal ratio
A = film speed (ASA)
B = brightness value (see table of photographic settings)

For example, taking a picture of prominences on the Sun during totality ($B = 50$) using a telephoto lens at f/11 with 400 ASA film would require an exposure of:

$$E = 11^2/(400 \times 50)$$
$$E = 121/20{,}000$$
$$E = 0.00605 \text{ second or } 1/165 \text{ second}$$

As normal 35mm cameras do not have an exposure of 1/165 second, use the closest shutter speeds, which would be 1/125 and 1/250 second. When possible, bracket your exposures on both sides of the ideal exposure and take several at the same settings to help ensure success. If you use a film with an ASA too high, you may discover that your camera does not have a fast enough shutter speed to allow the proper exposure. Calculate all of your camera settings before the eclipse so that you know in advance what film you will need with your equipment.

Eclipse Snapshots

In 1983, Leif Robinson, editor of *Sky & Telescope* magazine, tried an experiment that showed how anyone with a zoom lens can make a decent photograph of the totally eclipsed Sun with a hand-held camera. He was on Java and, like everyone else in his group, he had set up his camera. With five minutes left before totality, Robinson looked around and realized that he was competing with professional photographers. There would be little need for the pictures he would take. He wondered if he might be able to take a different kind of picture. In his pocket was a roll of the new Kodak 1000 film. As the light was fading, he tore open his camera, ripped out the film, and loaded the experimental film. He then held his camera in his hands while he took snapshots of totality. To his delight and surprise, the pictures turned out fine. The prints were a bit grainier than they would have been with slower films, but here at last was a way that anyone with a camera fitted with a telephoto lens could take a good picture of the Sun in total eclipse without a tripod or special equipment. The discovery was particularly good news for people aboard eclipse cruise ships who might not always have perfectly calm seas.

Eclipse Photography with a Telescope

A telephoto lens or telescope with a focal length of 400 mm or greater is recommended to give you close-up photographs of the Sun or Moon. However, because of the increased magnification and the fact that your platform is moving (the Earth is rotating on its axis), a sturdy equatorial mount with a clock drive will be required to obtain quality photographs through a telephoto lens or telescope. Much practice and experience are required before you attempt astrophotographs with a telescope even for something as bright as a solar eclipse.

Many modern telescopes are equipped with battery-driven clock drives, but some still use AC electricity. If your telescope drive uses DC power, bring along spare batteries. If it requires AC, check on a source of electricity at your observing site or use a drive corrector with a built-in converter that changes 12-volt DC to 60-cycle 110-volt AC.

Many countries do not have 60-cycle, 110-volt AC electricity. You should inquire as you make your travel plans and, if necessary, bring along an AC converter, available in many hardware stores at modest cost.

For your telescope to follow the Sun throughout the eclipse, the equatorial mount must be aligned with the celestial pole at the eclipse site. To ensure the best alignment, set up the telescope before sunrise while Polaris (or the south celestial pole) is still visible or align the telescope the night before and cover the equipment until it is needed. Or choose a position at your observing site such that the north or south celestial pole is marked by a distant terrestial feature visible in the daytime sky, for example, the top of a pole or antenna at which you can point your telescope to align it.

If you do not have the opportunity to align the telescope before sunrise, a magnetic compass is helpful in orienting the telescope to approximately true north. Take into account magnetic variation for your location, which can be determined from a topographical or aviation map of your observing site.

Video Photography

With the introduction of home video cameras, eclipse observers have a remarkable new tool with which to preserve the celestial event in a dynamic way. Since video cameras are automatic, you need not worry about shutter or film speeds; however, you will still need a solar filter in front of the camera lens during the partial phases of an eclipse. Pointing your video camera directly at the Sun without a solar filter can cause irreversible damage to the sensitive detector in the camera. *Be sure to remove the solar filter during totality.* To get the best results, place the camera on a sturdy tripod and use a telephoto lens to give you as large an image as possible.

Video photography also offers a chance to contribute useful scientific data. Recording an eclipse in conjunction with an accurate time signal allows the observer to determine the precise time of totality for a given location. This can be particularly important at the northern and southern limits of totality, where video recordings can provide the data needed for accurate measurements of the Sun's diameter. A time signal is broadcast 24 hours a day over shortwave radio station WWV, Fort Collins, Colorado, and from Hawaii on WWVH at 5, 10, and 15 MHz. Inexpensive radios capable of receiving time signals are available (such as Realistic Time Kube from Radio Shack) and operate on one 9-volt battery.

Another advantage of using a video camera to photograph a solar eclipse is that you can record the audio response of people, wind, and birds to the most beautiful and majestic celestial event visible from planet Earth.

Checklist for Solar Eclipse Photography

Camera (two, if possible, in case one fails)

Lenses (50mm f/2 to 2000mm f/10)

Tripod

Telescope (consult your owner's manual for photographic equipment) mount, wedge, cords, oculars, photo adapters (T-ring, etc.), counterweights, drive corrector (and a spare fuse), lens cleaning supplies

Solar filter (full aperture or off-axis)

Inexpensive Mylar filter for finder scope

Hand-held solar filter (for viewing partial phases)

Film (36-exposure for eclipse)

Cable release for camera

Extra batteries for camera

Flashlight

Small tape recorder (spare tapes and batteries)

Portable shortwave receiver (to record WWV time signal)

Exposure list

Extension cord (if necessary)

Battery to run telescope clock drive (if necessary)

Voltage corrector (if necessary)

Assortment of tools for minor repairs

Compass (to determine north)

Level (to level telescope mount)

Photographic Settings for a Solar Eclipse

$$E = f^2/(A \times B)$$

where
E = exposure (seconds)
f = focal ratio
A = film speed (ASA)
B = brightness value (B value)

Sun at full disk or partial eclipse
Through full aperture Solar-Skreen filter
B = ∼80

	(Film Speed—ASA)				
f/	**32**	**64**	**100**	**200**	**400**
2	1/500	1/1000	1/2000	«	«
2.8	1/250	1/500	1/1000	1/2000	«
4	1/125	1/250	1/500	1/1000	1/2000
5.6	1/60	1/125	1/250	1/500	1/1000
8	1/30	1/60	1/125	1/250	1/500
11	1/15	1/30	1/60	1/125	1/250
16	1/8	1/15	1/30	1/60	1/125
32	1/2	1/4	1/8	1/15	1/30

Sun at total eclipse: prominences
No filter
B = 50

	(Film Speed—ASA)				
f/	**32**	**64**	**100**	**200**	**400**
2	1/500	1/1000	1/1000	1/2000	«
2.8	1/250	1/500	1/500	1/1000	1/2000
4	1/125	1/250	1/250	1/500	1/1000
5.6	1/60	1/125	1/125	1/250	1/500
8	1/30	1/60	1/60	1/125	1/250
11	1/15	1/30	1/30	1/60	1/125
16	1/8	1/15	1/15	1/30	1/60
32	1/2	1/4	1/4	1/8	1/15

Sun at total eclipse: inner corona (3° field)
Through full aperture Solar-Skreen filter
B = 5

	(Film Speed—ASA)				
f/	**32**	**64**	**100**	**200**	**400**
2	1/30	1/60	1/125	1/250	1/500
2.8	1/15	1/30	1/60	1/125	1/250
4	1/8	1/15	1/30	1/60	1/125
5.6	1/4	1/8	1/15	1/30	1/60
8	1/2	1/4	1/8	1/15	1/30
11	1	1/2	1/4	1/8	1/15
16	3	1	1/2	1/4	1/8
32	15	6	4	2	1/2

Sun at total eclipse: outer corona (10° field)
No filter
B = 1

	(Film Speed—ASA)				
f/	**32**	**64**	**100**	**200**	**400**
2	1/8	1/15	1/30	1/60	1/125
2.8	1/4	1/8	1/15	1/30	1/60
4	1/2	1/4	1/8	1/15	1/30
5.6	1	1/2	1/4	1/8	1/15
8	4	2	1/2	1/4	1/8
11	8	3	2	1/2	1/4
16	20	9	5	2	1/2
32	135	50	30	12	5

[T]he general phaenomenon is perhaps the most awfully grand which man can witness.

George B. Airy (1851)

13 The Pedigree of an Eclipse

The eclipse family known as saros 136 was born on June 14, 1360, deep in the southern hemisphere, over Antarctica and the southern Indian Ocean. The Moon was at a descending node—crossing the Sun's path on its way south. The Sun just happened to be near that node at the time so that, as viewed from Earth, the Moon grazed the southwestern edge of the Sun. It was a very slight partial eclipse, unnoticeable to the eye without filters, and there was no one there to see.

It was an inauspicious birth, but the firstborn of every saros family is always a slight partial eclipse that brushes the Earth at one of the poles and gives no visual evidence of the splendor that will come as the family matures.

Celestial Clockwork

Time passed. The years rolled by. Forty-one other solar eclipses touched the Earth, but they came from other saros families. Then, after 6,585 days, the Moon had completed 223 lunations and the Sun had passed by the descending node of the Moon 19 times. The two cycles nearly matched, forcing the Sun and Moon to meet under almost the same circumstances that prevailed 18 years 11⅓ days earlier. Another eclipse was inevitable.

But the new eclipse was not identical to the first. The Moon's node was not exactly where it had been 18 years earlier. It had backed around the Sun's orbit and was now 0.477 degrees east of its previous position, so this time the Moon cut off a slightly larger portion of the Sun's light. It was still a very slight (0.199) partial eclipse, noticeable only with optical filters in the south Atlantic Ocean and Antarctica; and again no one was there to see.

Each 18 years and a few days brought a new solar eclipse of saros family 136, each a little farther north on the whole; each a partial eclipse, but each time covering a little more of the Sun. During 126 years there were eight partial eclipses, until on August 29, 1486, the Moon's diameter covered 98.6 percent of the Sun, leaving at maximum eclipse only a thin crescent of the Sun visible from near the south pole.

The next eclipse in the cycle, September 8, 1504, was different from those that preceded it. The Moon passed directly across the disk of the Sun as seen from near the Antarctic coast. It would have been a total eclipse if the angular size of the Moon's disk had been large enough, but the Moon's elliptical orbit had carried it farther from the Earth than average, so it appeared smaller—too small to cover the Sun completely. At maximum eclipse, the Sun's surface still shone around the circumference of the Moon to form a bright ring of light—an annular eclipse—lasting at most 31 seconds.

Throughout the sixteenth century, each of the six eclipses of saros 136 was an annular eclipse, gradually migrating northward. With each eclipse, the Moon was a little closer to Earth, so that the Moon's apparent size grew larger and the duration of the annulus got *shorter.* On November 12, 1594, the annular eclipse lasted 4 seconds at its midpoint. The disk of the Moon was almost large enough to completely hide the Sun.

At the next eclipse, it did—for just *1 second.* On November 22, 1612, in the southeastern Pacific Ocean and Antarctica, birds, fish, and whales saw an eclipse that was annular all along the central path until the eclipse reached its midpoint, where the surface of the Earth was closest to the Moon and hence the Moon appeared largest. At that location, the Moon's disk for just an instant completely covered the Sun and the eclipse (technically, at least) became total. The dark shadow of the Moon caused by saros 136 for the first time actually touched the Earth. Then, as modestly as a young boy and girl touch-

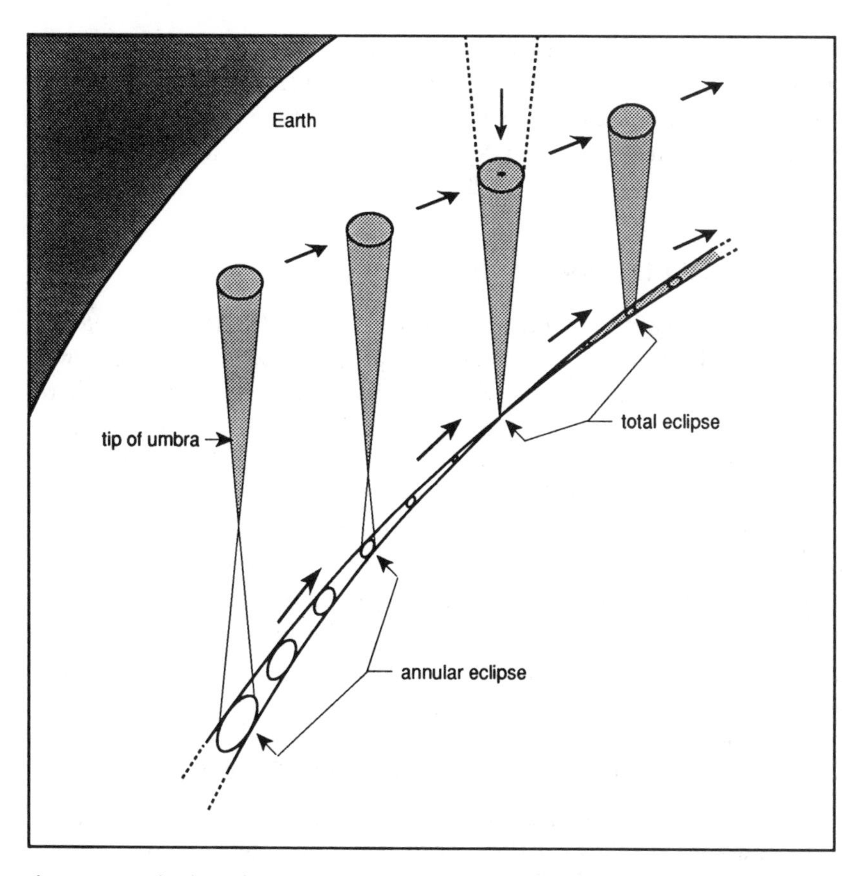

An occasional solar eclipse may start off annular but become total as the roundness of the Earth reaches up to intercept the shadow. The eclipse then returns to annular as the curvature of the Earth causes its surface to fall away from the shadow. These hybrids are called annular-total eclipses.

ing hands for the first time, the shadow and the Earth pulled apart as the Earth's surface curved away beneath the extended shadow. Just the slightly greater distance from the Moon that the Earth's curvature provided was enough to make the Moon once again appear too small to completely cover the Sun. The Sun peered out from around the entire circumference of the Moon, an annular eclipse once more.

The eclipse of 1612 and all five eclipses of saros 136 in the seventeenth century were hybrids of this kind: beginning as annular, becoming total, then returning to annular again. And with each eclipse, the duration of totality was a little longer, until on January 5,

1685, the total phase lasted 35 seconds and, at least momentarily, totality was seen for thousands of miles along the eclipse route. Only at the western and eastern extremes of the eclipse path, near sunrise and sunset, where the distance to the Moon from the surface of the Earth was slightly greater, was the eclipse annular.

The Big Leagues

On the twentieth eclipse of saros 136, the Earth was near its closest to the Sun, so the Sun's angular size was greatest and it was hardest to eclipse. But the Moon in its orbit was close enough to Earth so that its apparent size was larger than average. Thus, finally, on January 17, 1703, the eclipse was total all along the central path, with totality lasting a maximum of 50 seconds.

Each subsequent eclipse of saros 136 was total, and the duration of totality was rapidly increasing. By 1793 the eclipse lasted a maximum of 2 minutes 52 seconds over the Indian Ocean and provided a shorter show for the aborigines in Australia and the newly arrived British settlers. The eclipse path was shifting northward. As the

---

The Extinction of Total Solar Eclipses

In 1695, Edmond Halley discovered that eclipses recorded in ancient history did not match calculations for the times or places of those eclipses. Starting with records of eclipses in his day and the observed motion of the Moon and Sun, he used Isaac Newton's new theory of universal gravitation (1687) to calculate past eclipses and compare the results with records of eclipses actually observed more than 2,000 years earlier. They did not match. Halley had great confidence in the theory of gravitation and resisted the temptation to conclude that the force of gravity changed as time passed. Instead, he proposed that the length of a day on Earth must have increased by a small amount.

If the Earth's rotation had slowed down slightly, the Moon must have gained orbital velocity, because the Earth and Moon are a gravitational pair and their total momentum must be conserved. If the Earth lost momentum in its spin, the Moon must have gained momentum in its orbital speed. This acceleration of the Moon would have caused it to spiral slowly outward from the Earth, ascending to a more distant orbit, where it travels more slowly. If, 2,000 years earlier, the Earth was spinning a little faster and the

---

American Civil War ended, saros 136 crossed the 5-minute plateau into the realm of rare and great eclipses as it stretched from South America to southern Africa. The totality of April 25, 1865 lasted 5 minutes 23 seconds.

A total eclipse from saros 136 (June 30, 1973).
© 1973 Hans Vehrenberg

Moon was a little closer and orbiting a little faster, then eclipse theory and observation would match. Scientists soon realized that Halley was right.

The Earth's spin continues to slow because of tidal friction. The Moon is primarily responsible for the tides. As the shallow continental shelves (primarily in the Bering Sea) collide with high tides, the Earth's rotation is retarded. The slower spin of the Earth causes the Moon to gain speed and thus to move farther from the Earth. Astronomer Robert S. Harrington calculates that the Moon is receding from the Earth at a little less than 1 inch (about 2.3 cm) a year.

As the Moon recedes from Earth, its apparent disk becomes smaller, and the chances of an eclipse being total rather than annular are reduced. Total eclipses are moving toward extinction. When the Moon is 17,880 miles (28,770 km) farther from the Earth, its angular size will be so small that total eclipses will cease. With the Moon receding at almost 1 inch a year, the last total solar eclipse visible from the surface of the Earth will take place in 1.25 *billion* years. There is still time to catch one of these majestic events.

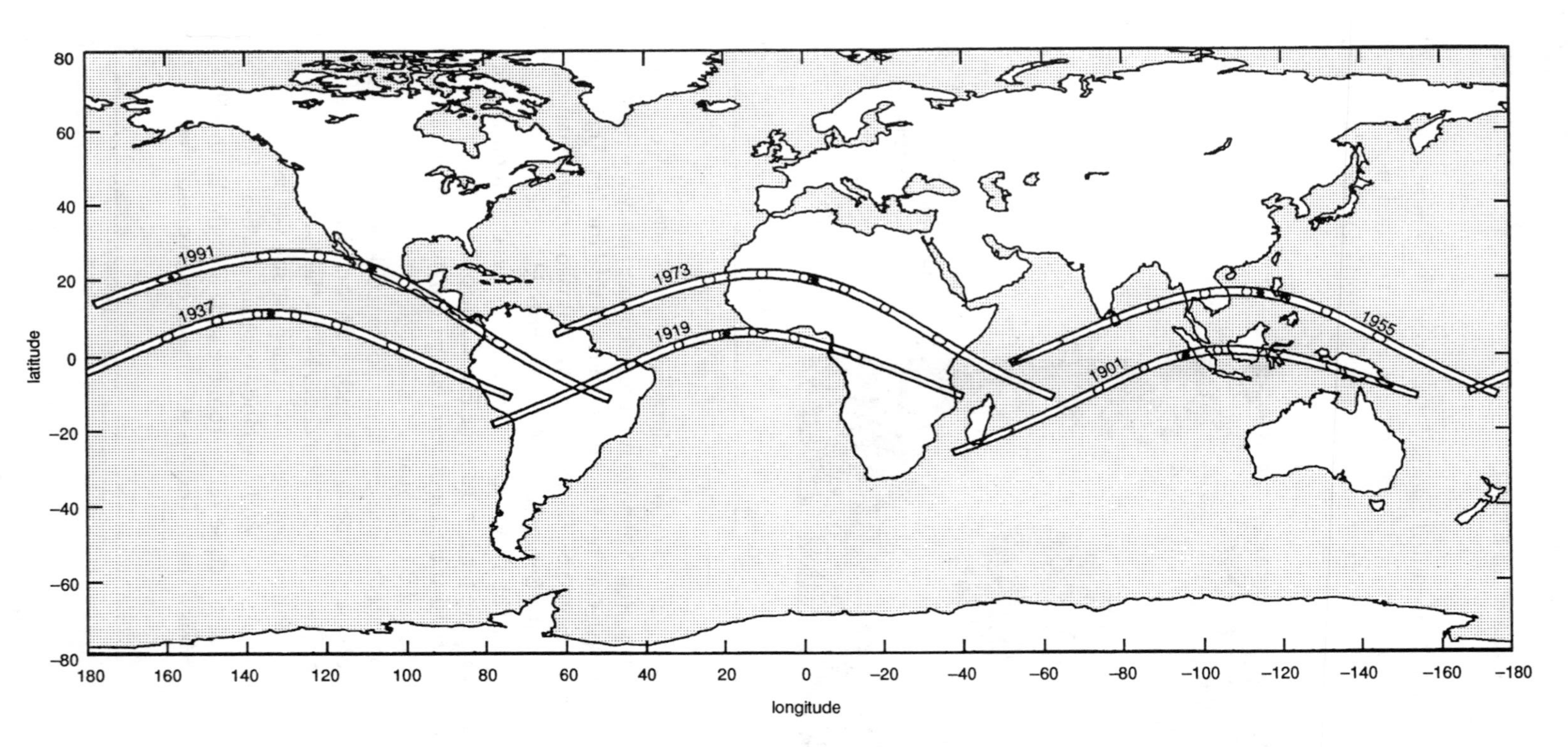

Paths of totality for the six eclipses of saros 136 that occur in the twentieth century. Successive eclipses shift westward and northward.

Now almost everything was conspiring to make saros 136 one of the greatest in history. The eclipses were occurring later and later in the spring, when the Sun was farther from Earth and hence smaller in apparent size, enabling total eclipses to last longer. At each eclipse, the perigee of the Moon (its position closest to Earth) was close to the eclipse point and getting closer, so that the Moon was near a maximum apparent size, allowing total eclipses to last longer. Every eclipse of saros 136 in the twentieth century would last longer than 6 minutes. And the twentieth century was prepared to make good use of this rare gift.

The eclipse of May 18, 1901, brought astronomers from all over the world to Indonesia to continue their analysis of the solar atmosphere, still best seen inside the shadow of the Moon. They had at most 6 minutes 29 seconds within which to work.

The next eclipse in the cycle lasted even longer—6 minutes 51 seconds at its peak, a totality long enough to make it memorable in and of itself. But that was not why eclipse number 32 in saros 136 became the most famous in history. Astronomers used that eclipse to measure star positions around the darkened Sun and concluded that starlight passing by the Sun had been bent, just as Einstein had predicted in his recently completed general theory of relativity. Thus the total eclipse of May 29, 1919, marked a major turning point in the history of science.

Saros 136, however, had its own dramatic schedule to fulfill, independent of the brilliance or depravity of man. It returned on June 8, 1937, as the clouds of war darkened over Europe and the Far East, with a 7 minute 4 second eclipse—the first to last over 7 minutes since 1098. Unfortunately, this amazing sight passed almost entirely across the expanse of the Pacific Ocean, avoiding land, where scientists could deploy their instruments.

Saros 136 reached its climax on June 20, 1955, with an eclipse that, at maximum, lasted 7 minutes 8 seconds, the longest since June 20, 1080 (7 minutes 18 seconds), and the longest until June 25, 2150 (7 minutes 14 seconds). The shadow visited Sri Lanka, Thailand, Indochina, and the Philippines. Never again would saros 136 equal that duration of totality. Yet few saroses even come close. June 30, 1973, brought an eclipse, visible across Africa, that lasted as long as 7 minutes 4 seconds.

The last offering of saros 136 in the twentieth century happens on July 11, 1991, bringing up to 6 minutes 54 seconds of totality and

A radial density filter was used to bring out coronal detail in this view from Kenya of a saros 136 eclipse (June 30, 1973). *HAO/NCAR*

tracing a path from Hawaii through Mexico, Central America, Colombia, and Brazil, visible to more people than any eclipse in history.

The six longest eclipses of the twentieth century were in 1901, 1919, 1937, 1955, 1973, and 1991—every one of them a member of the same eclipse family. Saros 136 is one of the greatest eclipse generation cycles in recorded history.

Closing Out a Career

Now, gradually, the glory is beginning to fade. Saros 136 is declining. The Moon is receding from perigee, making its disk appear smaller; it will not cover the Sun as well. Eclipses are occurring ever later in the year, when the Earth is actually closer to the Sun so that its disk appears larger and harder to eclipse. Eclipses will become steadily shorter, but they will still be total and will still be comparatively long for quite some time. The glory will fade slowly.

Saros 136 does not end in 1991. It returns on July 22, 2009, with a 6-minute-39-second display and a central path across India, China, and the western Pacific. It will be back on August 2, 2027, racing over North Africa, Saudi Arabia, and the Indian Ocean, bringing darkness for up to 6 minutes 23 seconds. It will pay its first visit to the continental United States on August 12, 2045, streaking from northern California through Florida, the longest totality for the continental United States in the calculated history of eclipses. Over the Caribbean, it will last for 6 minutes 6 seconds. But that will be the last eclipse of saros 136 to surpass the 6-minute mark.

Three saros cycles (54 years) later, on September 14, 2099, it will return to North America, slicing across southwestern Canada and plunging southeastward across the United States to the mouth of the Chesapeake Bay and off into the Atlantic. At maximum, east of the Windward Islands, the eclipse will last 5 minutes 18 seconds.

Saros 136 is aging, but its eclipses are still total and more than long enough to lure people by the thousands into its shadows. By May 13, 2496, however, a total eclipse of saros 136 will have dwindled to 1 minute 2 seconds for hardy travelers in the Arctic. It is the last of 45 total eclipses in this remarkable saros family. At the next visit of this saros, May 25, 2514, the dark shadow of the Moon will miss the Earth, passing above the north polar region. The remaining six eclipses in the sequence will all be partial as well, steadily declining in the Moon's coverage of the face of the Sun. On July 30, 2622, there will be one final partial eclipse, virtually unnoticeable near the north pole. Saros 136 will have died.

Saros number 136 will have performed on Earth for 1,262 years and created 71 solar eclipses. Not an exceptionally long career, but what a record! Of its 71 eclipses, 15 will have been partial, 6 annular, 5 a combination annular and total, and *45 total.* It brought the lucky people of the twentieth century three eclipses with totality exceeding 7 minutes and all six of the longest eclipses in that century. The typical saros offers an average of only 19 or 20 total eclipses. If this saros were an athlete, its shadow-black jersey with its corona-white number 136 would be retired to hang in glory in the Eclipse Hall of Fame.

The Eclipse Family of Saros 136

No.	Date	Type	Duration (minutes:seconds)
1	1360 June 14	p	[0.050]*
2	1378 June 25	p	[0.199]
3	1396 July 5	p	[0.346]
4	1414 July 17	p	[0.489]
5	1432 July 27	p	[0.626]
6	1450 August 7	p	[0.757]
7	1468 August 18	p	[0.876]
8	1486 August 29	p	[0.986]
9	1504 September 8	a	0:31
10	1522 September 19	a	0:23
11	1540 September 30	a	0:17
12	1558 October 11	a	0:12
13	1576 October 21	a	0:08
14	1594 November 12**	a	0:04
15	1612 November 22	a/t	totality 0:01
16	1630 December 4	a/t	totality 0:07
17	1648 December 14	a/t	totality 0:15
18	1666 December 25	a/t	totality 0:24
19	1685 January 5	a/t	totality 0:35
20	1703 January 17	t	0:50
21	1721 January 27	t	1:07
22	1739 February 8	t	1:28
23	1757 February 18	t	1:52
24	1775 March 1	t	2:20
25	1793 March 12	t	2:52
26	1811 March 24	t	3:27
27	1829 April 3	t	4:05
28	1847 April 15	t	4:44
29	1865 April 25	t	5:23
30	1883 May 6	t	5:58
31	1901 May 18	t	6:29
32	1919 May 29	t	6:51
33	1937June 8	t	7:04
34	1955 June 20	t	7:08
35	1973 June 30	t	7:04
36	1991 July 11	t	6:53
37	2009 July 22	t	6:39
38	2027 August 2	t	6:23

39	2045 August 12	t	6:06
40	2063 August 24	t	5:49
41	2081 September 3	t	5:33
42	2099 September 14	t	5:18
43	2117 September 26	t	5:04
44	2135 October 7	t	4:50
45	2153 October 17	t	4:36
46	2171 October 29	t	4:23
47	2189 November 8	t	4:10
48	2207 November 20	t	3:56
49	2225 December 1	t	3:43
50	2243 December 12	t	3:30
51	2261 December 22	t	3:17
52	2280 January 3	t	3:04
53	2298 January 13	t	2:52
54	2316 January 25	t	2:42
55	2334 February 5	t	2:32
56	2352 February 16	t	2:24
57	2370 February 27	t	2:17
58	2388 March 9	t	2:10
59	2406 March 20	t	2:03
60	2424 March 31	t	1:55
61	2442 April 11	t	1:46
62	2460 April 21	t	1:34
63	2478 May 3	t	1:21
64	2496 May 13	t	1:02
65	2514 May 25	p	[0.952]*
66	2532 June 5	p	[0.824]***
67	2550 June 16	p	[0.685]
68	2568 June 26	p	[0.544]
69	2586 July 7	p	[0.397]
70	2604 July 19	p	[0.252]
71	2622 July 30	p	[0.105]

p = partial a = annular t = total

*Magnitude of partial eclipse: fraction of Sun's diameter covered by the Moon.

**Henceforth on the Gregorian calendar, which replaced the Julian calendar in 1582 and dropped 10 days from that year to keep March at the beginning of spring.

***The following magnitudes were provided by Fred Espenak.

Note: Saroses with even numbers occur at the descending node; they start near the south pole and progress northward. Saroses with odd numbers occur at the ascending node; they start near the north pole and progress southward.

Saros Series Statistics

	Range	Average
Number of solar eclipses in a series	70–85	73
Timespan for a series	1,244–1,514 years	1,315 years

At any time, 42 saros series are running simultaneously.

[A] total eclipse of the Sun . . . is the most sublime and awe-inspiring sight that nature affords.
Isabel Martin Lewis (1924)

14

The Eclipse of July 11, 1991

The Moon's shadow has darkened space for 4½ billion years, since the Sun formed and began to shine and the planets and their moons formed and could not shine. From the newly formed Sun, light streamed outward in all directions. Here and there it illuminated a body of rock or gas or ice. The Sun's dominant light identified that world as one of its own children.

A little over eight minutes of light-travel time outbound from the Sun, a portion of the light encountered two small dark bodies and bathed the sunward half of each in brightness. The surrounding flood of unimpeded light sped on, leaving a cone of darkness—a shadow—behind the planet and its one large moon. The two worlds lay the same average distance from the Sun. Not long before—measured on a cosmic timescale—they had been a single body, but they were separate and utterly different now.[1]

A Cosmic Birth

The Sun, planets, moons, comets, and asteroids had all begun within a cloud of gas and dust, a cloud so large that there was material enough to make dozens or hundreds of solar systems. Where the density was great enough, fragments of the cloud began to condense by gravity. At the heart of each fragment, a star (or two or more) was coalescing. Near those stars-to-be, other bodies, too low in mass ever to reach starhood, also began to form. These small bodies were planetesimals, the beginnings of the planets. At first they grew by gentle collisions and adhesions, gathering up a grain of ice, a fleck of dust along their paths around the Sun until icy or rocky planetesimals had taken shape. And still they gathered dust and small debris until they were so massive that gravity became their prime means of growth, gathering to them still more materials, and other planetesimals. The number of small bodies declined. The size of a few large bodies increased. The planets had been formed by convergence.

Convergence brought near disaster as well. Another planet-size body (the size of Mars today) wandered into the path of a body that would one day be called the Earth. No living thing witnessed the collision. No living thing could have survived the collision that obliterated the smaller world and nearly shattered the Earth. The Earth recoiled from the impact by spewing molten fragments of its crust and mantle outward. Some escaped from Earth; others rained down from the skies, pelting the surface in a rock storm of unimaginable proportions. But many of the fragments, caught in the Earth's gravity, stayed aloft, orbiting the Earth as the Earth orbited the Sun. Gradually, by collisions and accretion, the fragments joined together to form a new world circling the first. That new world was the Moon.

From convergence had come divergence. The two worlds, sprung from one, continued to diverge. Both were the same distance from the Sun, but the Earth was 81 times more massive than the Moon. That mass allowed the Earth to hold an atmosphere by gravity, while the Moon could not.

The eons passed. Life arose on Earth and covered the planet. Plants and animals responded to the tides raised by the Moon. The lunar tides slowed the Earth and caused the Moon to spiral slowly outward, diverging ever farther from the Earth in distance and ever further from the Earth in environment as well.

The lifeless Moon withdrew until today its shadow can just barely reach one natural body on a regular basis. As shadows in the universe go, this one is of no great size: a cone of darkness extending at most only 236,000 miles (379,700 km) in length before dwindling to a point. It is long enough to touch the Earth only occasionally and very briefly and with a single narrow stroke.

The black insubstantial cone reaches out but, for most of the time, there is nothing to touch. The shadow sweeps on through space unseen, unnoticed.

Contact

Ahead lies the Earth. A point of darkness collides with a world of rock and water and air. Suddenly the shadow becomes visible. Like a black winged spirit, the shadow swoops in from the heavens, silently

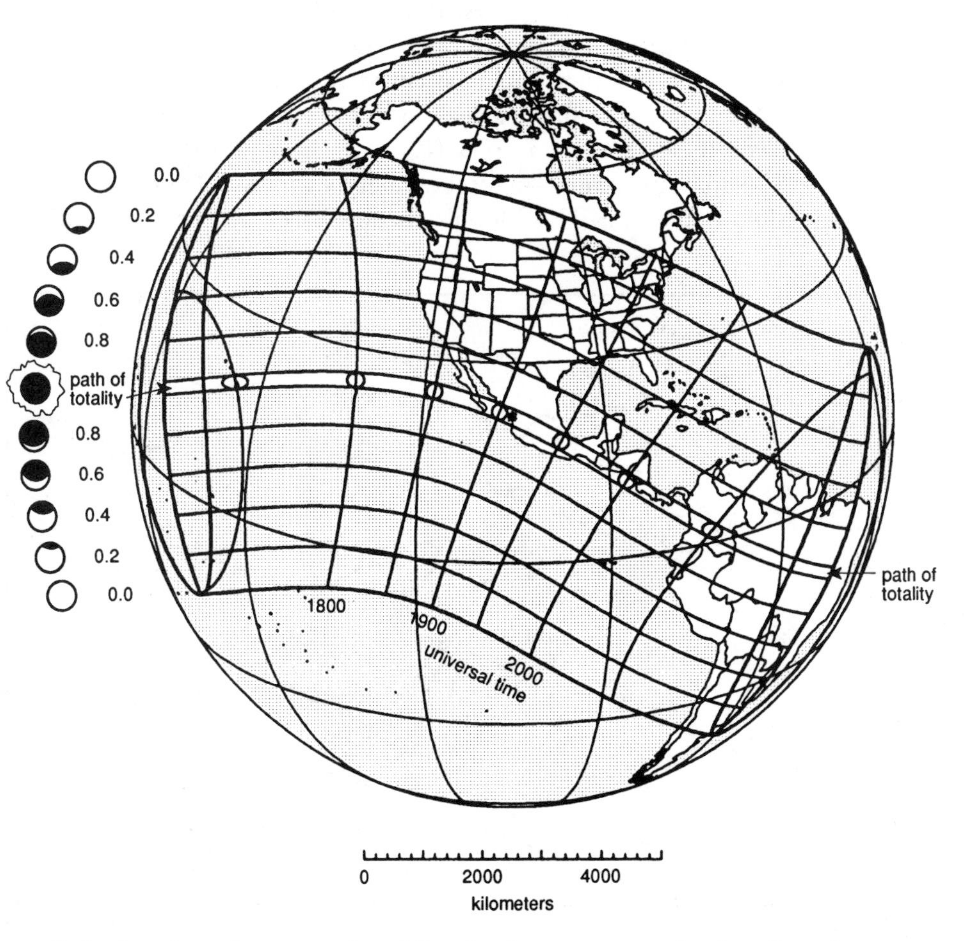

Path of totality and contours of partiality, showing percentages, for the eclipse of July 11, 1991. *Maps in this chapter by Fred Espenak, NASA Goddard Space Flight Center*

darkening the sky as it alights upon the surface of the Earth, on this occasion upon the Pacific Ocean, skimming the waters at a speed unknown to any aircraft. The date is July 11, 1991.

The shadow strikes the Earth as it always does—as it must—at sunrise. On this occasion it begins its ceaseless rush across the Earth's surface about 1,200 miles (1,900 km) west-southwest of the island of Hawaii.

Seen from space, the shadow sweeps eastward across the Pacific like a giant black felt-tip marker whose ink vanishes as it moves. The shadow makes landfall for the first time on this journey 4½ minutes after it touched down. It slides ashore in the middle of the west coast of Hawaii, an island that could be called the tallest mountain on Earth. Mauna Kea, the highest peak in this family of five active and extinct volcanoes, soars 6.3 miles (10.2 km) from the floor of the Pacific Ocean to its summit—a total rise nearly a mile (1.3 km) greater than Mount Everest.[2] The Moon's shadow races up the slopes of

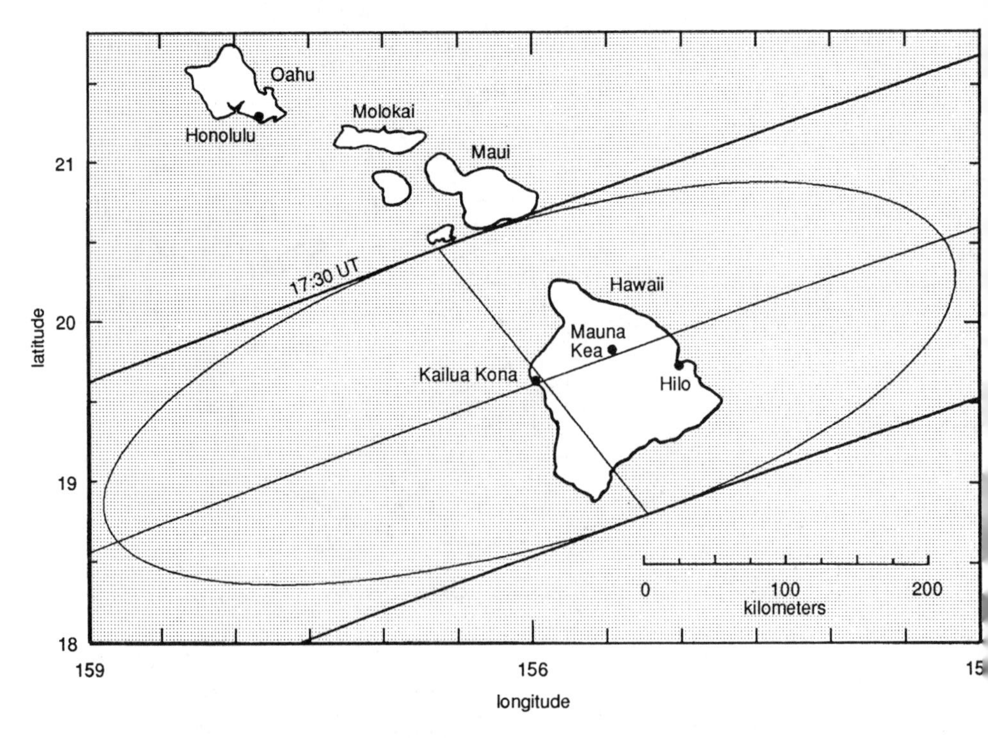

Path of totality over the island of Hawaii, July 11, 1991

Mauna Kea, and the center of the shadow passes within 1.5 miles (2.4 km) of the volcano's summit, home of the largest complex of astronomical observatories on Earth.

From all over the world, people have come to Hawaii to stand in the shadow of the Moon, to watch the blinding bright face of the Sun vanish in broad daylight and reveal its feathery white outer atmosphere (the corona) and the scarlet filaments (prominences) that hover at the base of the corona, just above the rim of the Moon.

Now and only now, in a total eclipse of the Sun, are these features revealed to the creatures of Earth without need of highly specialized equipment. Time after time, experienced professional astronomers, veterans of eclipse pilgrimages, have forgotten their research assignments and stood frozen in awe at what they proclaim to be the single most beautiful, most unforgettable sight in all the heavens.

Nearly every human being on the island has been up before dawn, talking excitedly, preparing anxiously, watching as the shadow nears and the Sun is eaten away, cookie-like, bite by bite, by the occulting Moon. Submerged completely in the shadow of the Moon, they stand hushed before the actual sight, except perhaps for an unconscious murmur of wonder. The hush lasts—the shadow lasts—up to 4 minutes 10 seconds on Hawaii.

Ever Onward

The shadow moves on, rushing eastward across the Pacific for a landfall on the North American continent. One hour 10 minutes after it leaves Hawaii, the shadow arrives in Mexico at the southern tip of Baja California. At the end of the Baja California peninsula the local time is near noon and the Sun is nearly overhead. The shadow on the ground is wider now—160 miles (258 km) across—only 7 miles (11 km) less than the theoretical maximum. The Earth is just 5 days past aphelion, its greatest distance from the Sun, so the Sun's apparent disk is near minimum angular size and easiest to cover fully. Meanwhile, the Moon is just 5½ hours past perigee, its least distance from Earth, so its apparent disk is near maximum size. It is 8 percent larger in angular size than the Sun. This substantial overlap creates a very long totality. Because the shadow is so wide, both the sky and ground around observers are darker than usual during totality.

Baja California has never received more visitors. Telescopes and

Local Contact Times for the Total Eclipse of July 11, 1991

[Local time difference from UT (Greenwich time) shown in parentheses after location]

*indicates site is along center line of eclipse

Site	Contact				Duration of Totality (minutes: seconds)
	1st	2nd	3rd	4th	
Hawaii (Big Island)					
Mauna Kea* (UT minus 10)	6:31 A.M.	7:28 A.M.	7:32 A.M.	8:38 A.M.	4:12
Mexico					
Baja California: 30 miles (50 km) south or southeast of La Paz* (UT minus 7)	10:23 A.M.	11:48 A.M.	11:54 A.M.	1:19 P.M.	6:48
Mazatlán (UT minus 7)	10:33 A.M.	11:59 A.M.	12:04 P.M.	1:28 P.M.	5:27
Tuxpan (on coast 110 miles [175 km] southeast of Mazatlán)* (UT minus 7)	10:36 A.M.	12:02 P.M.	12:09 P.M.	1:32 P.M.	6:52
Guadalajara (UT minus 6)	11:42 A.M.	1:09 P.M.	1:15 P.M.	2:38 P.M.	6:14
Morelia (UT minus 6)	11:48 A.M.	1:16 P.M.	1:22 P.M.	2:44 P.M.	6:20
Toluca* (UT minus 6)	11:53 A.M.	1:20 P.M.	1:26 P.M.	2:47 P.M.	6:42
Mexico City (UT minus 6)	11:54 A.M.	1:21 P.M.	1:28 P.M.	2:47 P.M.	6:34
Cuernavaca (UT minus 6)	11:54 A.M.	1:21 P.M.	1:28 P.M.	2:48 P.M.	6:32
Puebla (UT minus 6)	11:57 A.M.	1:24 P.M.	1:30 P.M.	2:50 P.M.	6:24

Oaxaca (UT minus 6)	12:03 P.M.	1:30 P.M.	1:36 P.M.	2:55 P.M.	5:25
Salina Cruz (UT minus 6)	12:08 P.M.	1:35 P.M.	1:41 P.M.	2:59 P.M.	5:26
Tapachula* (UT minus 6)	12:18 P.M.	1:42 P.M.	1:48 P.M.	3:04 P.M.	6:16
Central America					
Guatemala City, Guatemala (UT minus 6)	12:23 P.M.	1:47 P.M.	1:52 P.M.	3:07 P.M.	5:20
San Salvador, El Salvador (UT minus 6)	12:27 P.M.	1:51 P.M.	1:56 P.M.	3:10 P.M.	5:05
Managua, Nicaragua (UT minus 6)	12:37 P.M.	2:00 P.M.	2:02 P.M.	3:15 P.M.	2:45
San José, Costa Rica (UT minus 6)	12:45 P.M.	2:05 P.M.	2:10 P.M.	3:10 P.M.	5:07
David, Panama (UT minus 5)	1:51 P.M.	3:09 P.M.	3:15 P.M.	4:23 P.M.	5:25
Colombia (UT minus 5)					
Buenaventura	2:08 P.M.	3:22 P.M.	3:27 P.M.	4:31 P.M.	4:45
Cali	2:10 P.M.	3:23 P.M.	3:28 P.M.	4:32 P.M.	4:39
Neiva	2:12 P.M.	3:25 P.M.	3:30 P.M.	4:33 P.M.	4:51
Brazil					
Manicoré (UT minus 4)	3:40 P.M.	4:42 P.M.	4:46 P.M.	5:41 P.M.	3:53
Peixe (UT minus 3)	4:51 P.M.	5:47 P.M.	5:50 P.M.	(Sun has set)	3:16

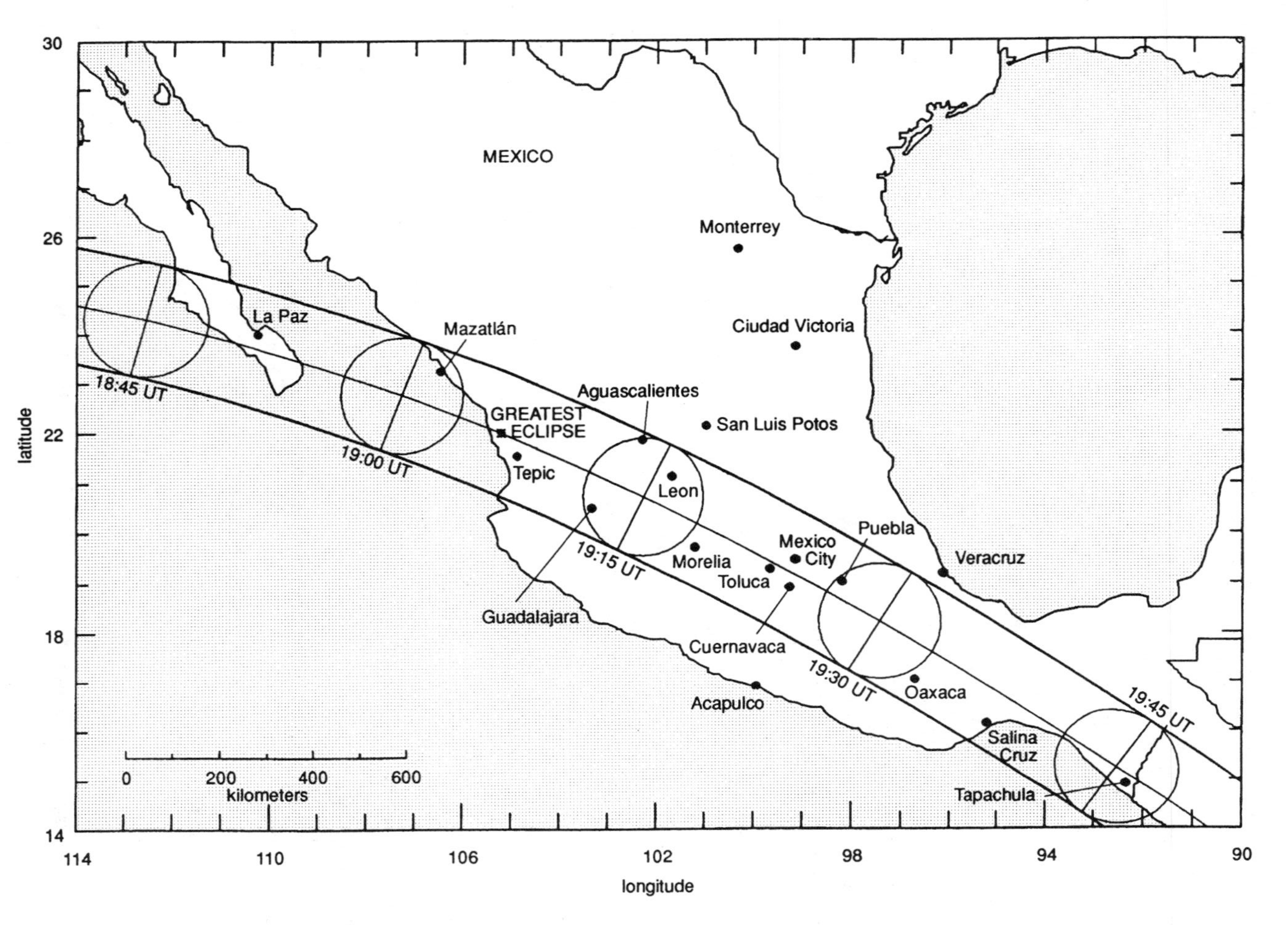

Path of totality over Mexico, July 11, 1991

cameras stand like soldiers at attention, but with eyes upward. Except to adjust alignment or reload film, the people stand transfixed beside their instruments, like them staring skyward. Here totality lasts 6 minutes 52 seconds.

Because the Moon's apparent size is so large, it blocks from view during the middle of totality all but the tallest prominences along the limb of the Sun. At this eclipse, the prominences are best seen at second contact, when the Moon is just covering the eastern edge of the Sun, and at third contact, as the Moon begins to uncover the western edge of the Sun.

The shadow, rushing onward, rides the wavetops in a slanting course across the Gulf of California. Enroute, the duration of totality reaches its maximum for this eclipse: 6 minutes 54 seconds. Not for 142 years, until June 13, 2132, will the Moon's shadow linger longer over any site on Earth.

The shadow enters mainland Mexico 50 miles (80 km) south of Mazatlán. Just ashore, the Sun stands at the zenith with the Moon covering its face and the cone of the lunar shadow aimed most nearly at the center of the Earth.[3] This, technically, is *eclipse maximum*, when the centers of the Sun, Moon, and Earth are most closely aligned. The duration of totality is still near its longest: 6 minutes 53 seconds. Here the eclipse shadow is moving the slowest it will travel on this passage, 1,450 miles (2,330 km) per hour.

The Moon, in its journey across the face of the Sun, is headed slightly southward, and as the curve of the Earth slopes away the shadow steers steadily more southeast. "Guadalajara, Mexico City, Cuernavaca, Puebla, Oaxaca," announces a map of the eclipse path, but this shadow express never stops. In Mexico City 21 million people experience 6 minutes 34 seconds of early afternoon darkness. Just outside the capital of Mexico, this mountain-climbing eclipse soars up and over the 18,700-foot (5,700-m) volcano Citlaltépetl, the highest peak in Mexico.

Onward, southeastward, the shadow goes, gathering speed as it begins a slide toward the edge of the Earth. The shadow has spent an hour visiting Mexico. Over the next 30 minutes, its southeastward course serendipitously tracks the thin spine of Central America. It skirts the Pacific coasts of Guatemala, El Salvador, Honduras, Nicaragua, Costa Rica, Panama.

The sunspot cycle reached its most recent maximum in 1990, so

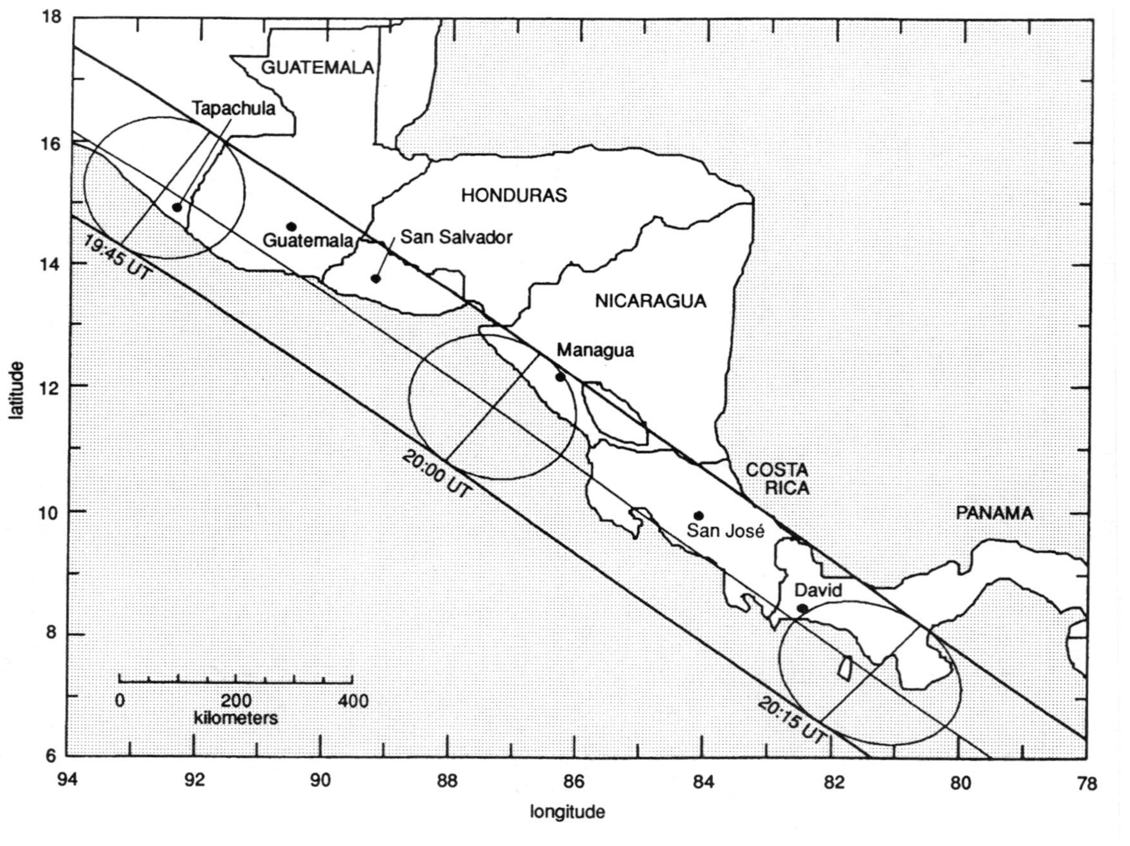

Path of totality along the west coast of Central America, July 11, 1991

the corona glows bright and round in shape, and there are many prominences to be seen.

Departure

Next the shadow enters South America, midway along the Pacific coast of Colombia. This mountaineering shadow rollercoasters over the three spines of the Cordillera. The middle summit is Nevado del Huila, reaching 18,865 feet (5,750 m) into the sky. Here midway across the Andes, the shadow crosses the highest mountain along its

---

Equipment Checklist for Eclipse Day

List of your intended activities during eclipse

Camera Equipment

Binoculars and/or telescope

Solar filters for eyes

Solar filters for binoculars and/or telescope

Thin sheets of cardboard (or a box) prepared to be a pinhole camera for indirect viewing of partial phases

Portable seat or ground covering

Flashlight with new batteries

Pencil and paper to record impressions or to sketch (also to take down the names and addresses of fellow observers)

Suitable clothing and hat (you will be outside for several hours)

Sunglasses (*not* for viewing partial phases)

Bug repellent, sunscreen lotion, basic first aid kit

Snacks and a canteen of water

Tape recorder into which you can quietly dictate your impressions of the eclipse or those of people near you or the reaction of wildlife

Tape recorder with earphones and prerecorded tape timed to cue you about what you want to do next (to run from about 2 minutes before totality until 2 minutes after totality)

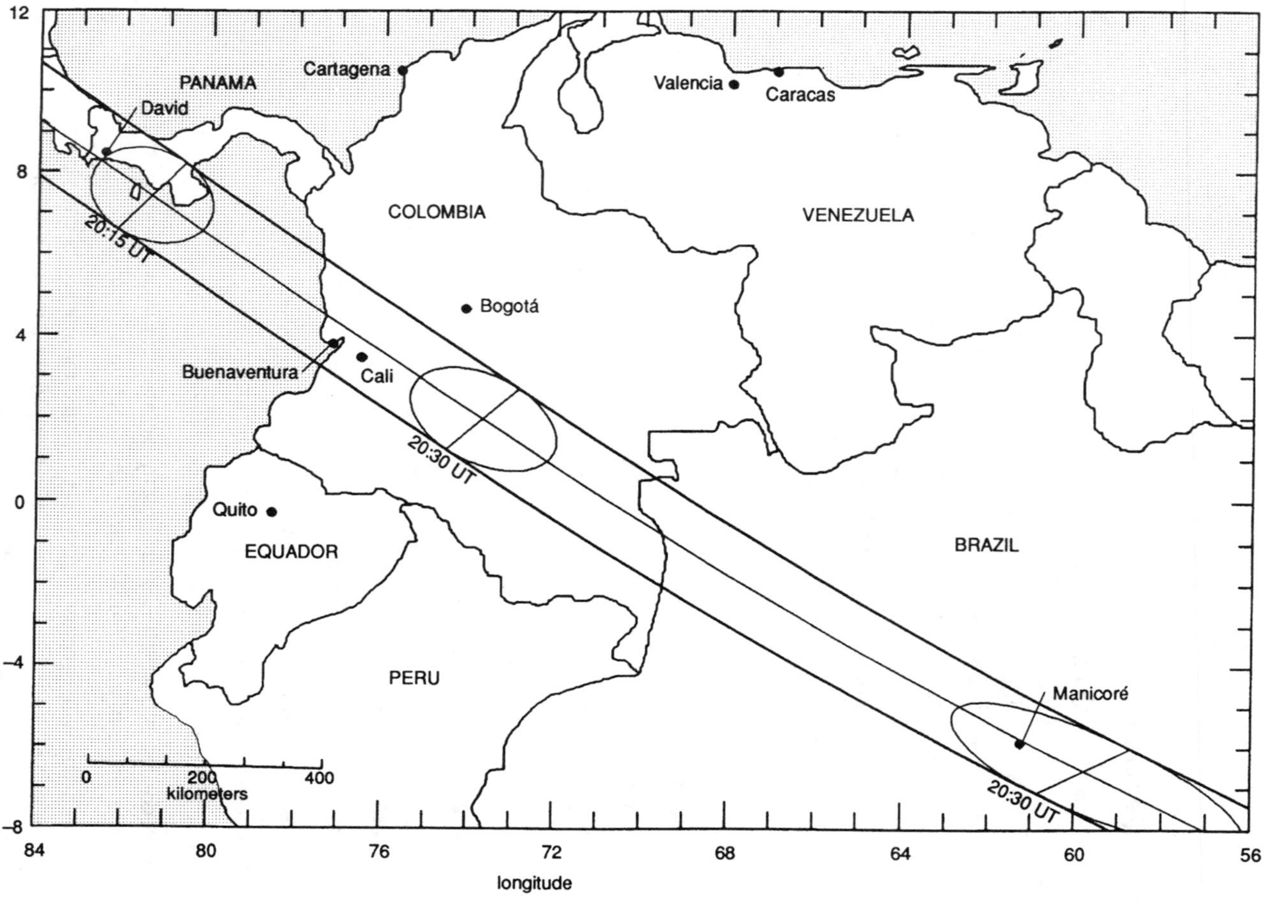

Path of totality over Colombia and Brazil, July 11, 1991

path. The darkness cascades down, then up, then down the eastern side of the Andes and into the Amazon Basin, across Colombia in 10 minutes and flows on into Brazil—the shadow width narrowing, the shadow speed increasing, the total eclipse time shortening to less than 4 minutes, late in the Brazilian afternoon.

The shadow rushes eastward, away from the Sun setting in the west, gathering speed like some giant black bird racing across the land, preparing to leap into the sky. Suddenly it is over. The shadow has left the Earth. It lifted off, vanishing into space, at sunset in Brazil about 200 miles (320 km) north-northeast of Brasília.

Gone. This encounter with a shadow is over. The Moon's

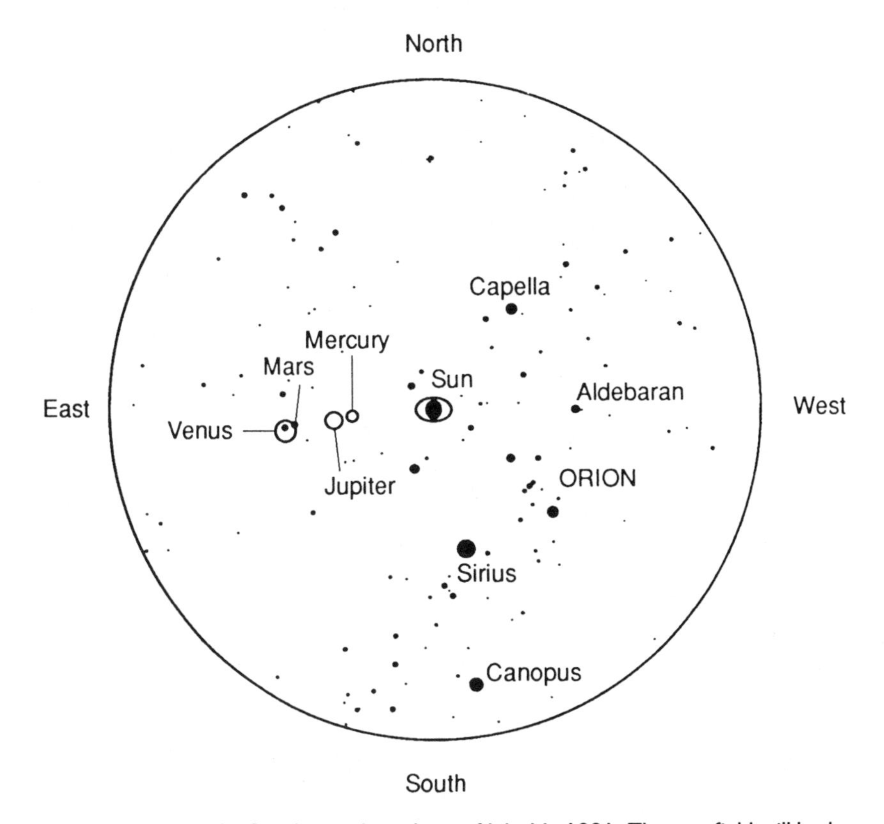

Star field around the Sun during the eclipse of July 11, 1991. The star field will look about the same from all locations along the path of totality, but the horizon will obscure parts of the eastern portion of the sky (including the planets east of the Sun) for viewers in Hawaii and parts of the western sky for viewers in Central and South America.

shadow will brush the Earth again in the eclipse seasons ahead. But there will never be an eclipse exactly like this one. Each is unique.

This one grazed and graced the Earth for 3 hours 26 minutes along a path 9,300 miles (15,000 km) long. The shadow drew its curving line across less than 1 percent of the Earth's surface, but touched 48 million people, the most ever covered by the shadow of the Moon, the most ever to have seen the Sun extinguished in the midst of the day.[4]

A total eclipse of the Sun. Special are the people privileged to witness this event and feel its power. The shadow may vanish but its memory is indelible.

Appendix A

Coming Attractions

Total Eclipses, 1991–2052

Date	*Maximum Duration of Totality (minutes: seconds)*	*Path of Totality*
1991, July 11	6:54	Hawaii, Mexico, Central America, Colombia, Brazil
1992, June 30	5:21	Uruguay, South Atlantic Ocean
1994, November 3	4:24	Peru, Chile, Bolivia, Paraguay, Brazil, South Atlantic Ocean
1995, October 24	2:10	Iran, Pakistan, India, Bangladesh, Burma, Cambodia, Vietnam, Pacific Ocean
1997, March 9	2:51	Mongolia, U.S.S.R.
1998, February 26	4:09	Pacific Ocean, Colombia, Panama, Venezuela, Atlantic Ocean, western Africa
1999, August 11	2:23	Atlantic Ocean, England, France, Germany, Austria, Hungary, Rumania, Turkey, Iraq, Iran, Pakistan, India
2001, June 21	4:57	Atlantic Ocean, Angola, Zambia, Rhodesia, Mozambique, Madagascar
2002, December 4	2:04	Angola, Botswana, Rhodesia, South Africa, Mozambique, south Indian Ocean, southern Australia
2003, November 23	1:59	Antarctica
2005, April 8	0:42	Pacific Ocean (annular-total)
2006, March 29	4:07	Eastern Brazil, Atlantic Ocean, Ghana, Togo, Dahomey, Nigeria,

Date	*Maximum Duration of Totality (minutes: seconds)*	*Path of Totality*
		Niger, Chad, Libya, Egypt, Turkey, U.S.S.R.
2008, August 1	2:28	Greenland, Arctic Ocean, U.S.S.R., Mongolia, China
2009, July 22	6:39	India, China, Pacific Ocean
2010, July 11	5:20	South Pacific Ocean, Chile, Argentina
2012, November 13	4:03	Northern Australia, South Pacific Ocean
2013, November 3	1:40	Atlantic Ocean, equatorial Africa (annular-total)
2015, March 20	2:47	North Atlantic and Arctic oceans
2016, March 9	4:10	Indonesia, Pacific Ocean
2017, August 21	2:40	Pacific Ocean, United States (Oregon, Idaho, Wyoming, Nebraska, Missouri, Kentucky, Tennessee, South Carolina), Atlantic Ocean
2019, July 2	4:33	South Pacific Ocean, Chile, Argentina
2020, December 14	2:10	Pacific Ocean, Chile, Argentina, South Atlantic Ocean, South Africa
2021, December 4	1:55	Antarctica
2023, April 20	1:16	South Indian Ocean, New Guinea, Pacific Ocean (annular-total)
2024, April 8	4:28	Pacific Ocean, Mexico, United States (Texas, Oklahoma, Arkansas, Missouri, Illinois, Indiana, Ohio, Pennsylvania, New York, Vermont, New Hampshire, Maine), southeastern Canada, Atlantic Ocean
2026, August 12	2:19	Greenland, Iceland, Spain

Date	*Maximum Duration of Totality (minutes: seconds)*	*Path of Totality*
2027, August 2	6:23	Atlantic Ocean, Morocco, Spain, Algeria, Libya, Egypt, Saudi Arabia, Yemen, Somalia
2028, July 22	5:10	South Indian Ocean, Australia, New Zealand
2030, November 25	3:44	South West Africa, Botswana, South Africa, south Indian Ocean, southeastern Australia
2031, November 14	1:08	Pacific Ocean (annular-total)
2033, March 30	2:38	Alaska, Arctic Ocean
2034, March 20	4:09	Atlantic Ocean, Nigeria, Cameroon, Chad, Sudan, Egypt, Saudi Arabia, Iran, Afghanistan, Pakistan, India, China
2035, September 2	2:55	China, Korea, Japan, Pacific Ocean
2037, July 13	3:58	Australia, New Zealand
2038, December 26	2:19	South Indian Ocean, southern Australia, New Zealand, Pacific Ocean
2039, December 15	1:52	Antarctica
2041, April 30	1:51	South Atlantic Ocean, Angola, Zaire, Uganda, Kenya, Somalia
2042, April 20	4:51	Indonesia, Philippines, Pacific Ocean
2043, April 9	0:00	Northeastern U.S.S.R. (marginal)
2044, August 23	2:05	Greenland, Canada, United States (Montana, North Dakota)
2045, August 12	6:06	United States (California, Nevada, Utah, Colorado, Kansas, Oklahoma, Arkansas, Mississippi, Alabama, Georgia, Florida), Haiti, Dominican Republic, Venezuela,

Date	*Maximum Duration of Totality (minutes: seconds)*	*Path of Totality*
		Guyana, Surinam, French Guiana, Brazil
2046, August 2	4:51	Eastern Brazil, Atlantic Ocean, southern Africa, south Indian Ocean
2048, December 5	3:28	South Pacific Ocean, Chile, Argentina, south Atlantic Ocean, South West Africa
2049, November 25	0:38	Indian Ocean, Indonesia (annular-total)
2050, May 20	0:22	South Pacific Ocean (annular total)
2052, March 30	4:08	Pacific Ocean, Mexico, United States (Florida, Georgia, South Carolina), Atlantic Ocean

Based on Hermann Mucke and Jean Meeus: *Canon of Solar Eclipses: –2003 to +2526* (Vienna: Astronomisches Büro, 1983) and Fred Espenak: *Fifty Year Canon of Solar Eclipses: 1986–2035* (Cambridge, Massachusetts: Sky Publishing, 1987; NASA Reference Publication 1178 revised).

Total and Annular Eclipses Visible from the United States, 1990–2052

Date	*Total or Annular*	*Path*
1991, July 11	t	Hawaii, Mexico, Central America, northern South America
1992, January 4	a	Pacific Ocean; ends at California coast
1994, May 10	a	Southwestern U.S. through Midwest to northeastern states
2012, May 20	a	China and Japan through California to Texas
2017, August 21	t	Oregon through Missouri and Georgia
2023, October 14	a	Oregon through New Mexico, Mexico, Central America, northern South America
2024, April 8	t	Mexico through Texas and Maine
2044, August 23	t	Greenland, western Canada, Montana
2045, August 12	t	California through Arkansas, Florida, Haiti, and northeastern South America
2046, February 5	a	Indonesia, Pacific Ocean, California, Nevada
2048, June 11	a	Oklahoma through Illinois, Canada, Scandinavia, U.S.S.R.
2052, March 30	t	Pacific Ocean, Mexico, northern Florida

Appendix B maps courtesy of Jean Meeus and the publisher from *Canon of Solar Eclipses,* by Jean Meeus, Carl C. Grosjean, and Willy Vanderleen (Oxford: Pergamon Press, 1966).

Hawaii has been plotted 10 degrees too far east on these maps.

Appendix B

Maps of Eclipse Paths, 1986–2052

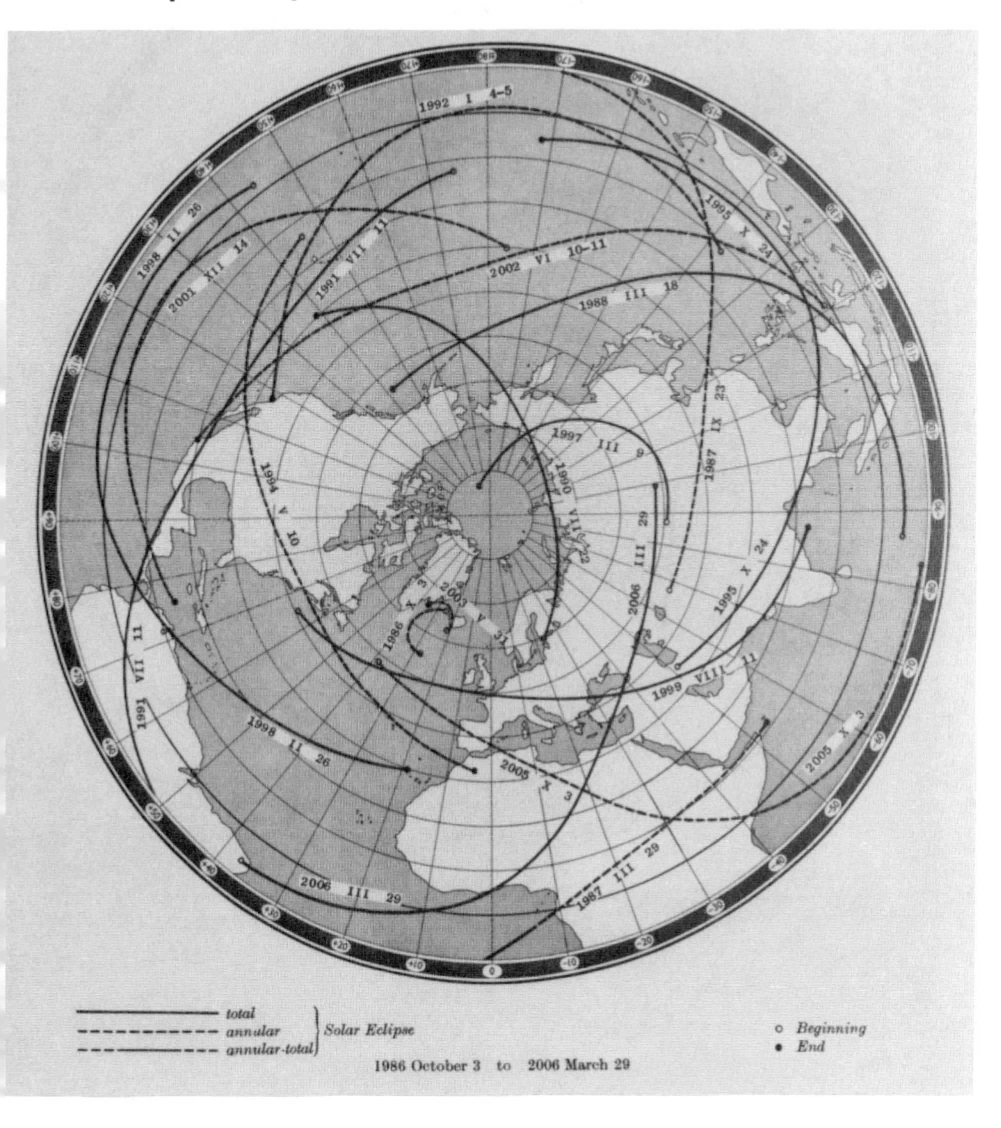

1986 October 3 to 2006 March 29

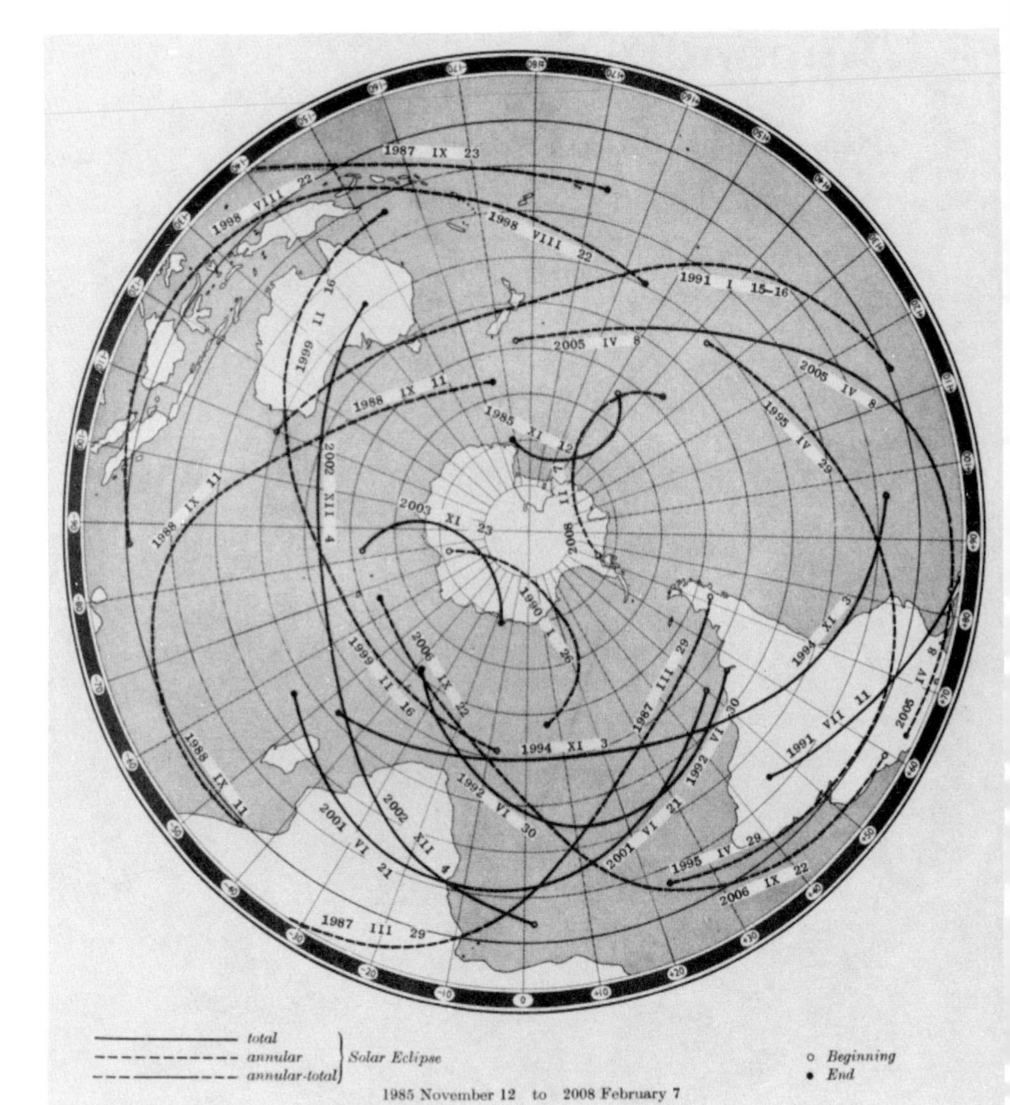
1987 IX 23
1998 VIII 22
1998 VIII 22
1991 I 15–16
1999 II 16
2005 IV 8
1988 IX 11
2005 IV 8
1985 XI 12
1995 IV 29
2002 XII 4
1988 IX 11
2003 XI 23
2008 II 7
1990 I 26
1994 XI 3
1999 II 16
2006 IX 22
1987 III 29
1992 VI 30
2005 IV 8
1991 VII 11
1994 XI 3
1988 IX 11
1992 VI 30
2002 XII 4
2001 VI 21
2001 VI 21
1995 IV 29
2006 IX 22
1987 III 29
total
annular
annular-total
Solar Eclipse
Beginning
End
1985 November 12 to 2008 February 7

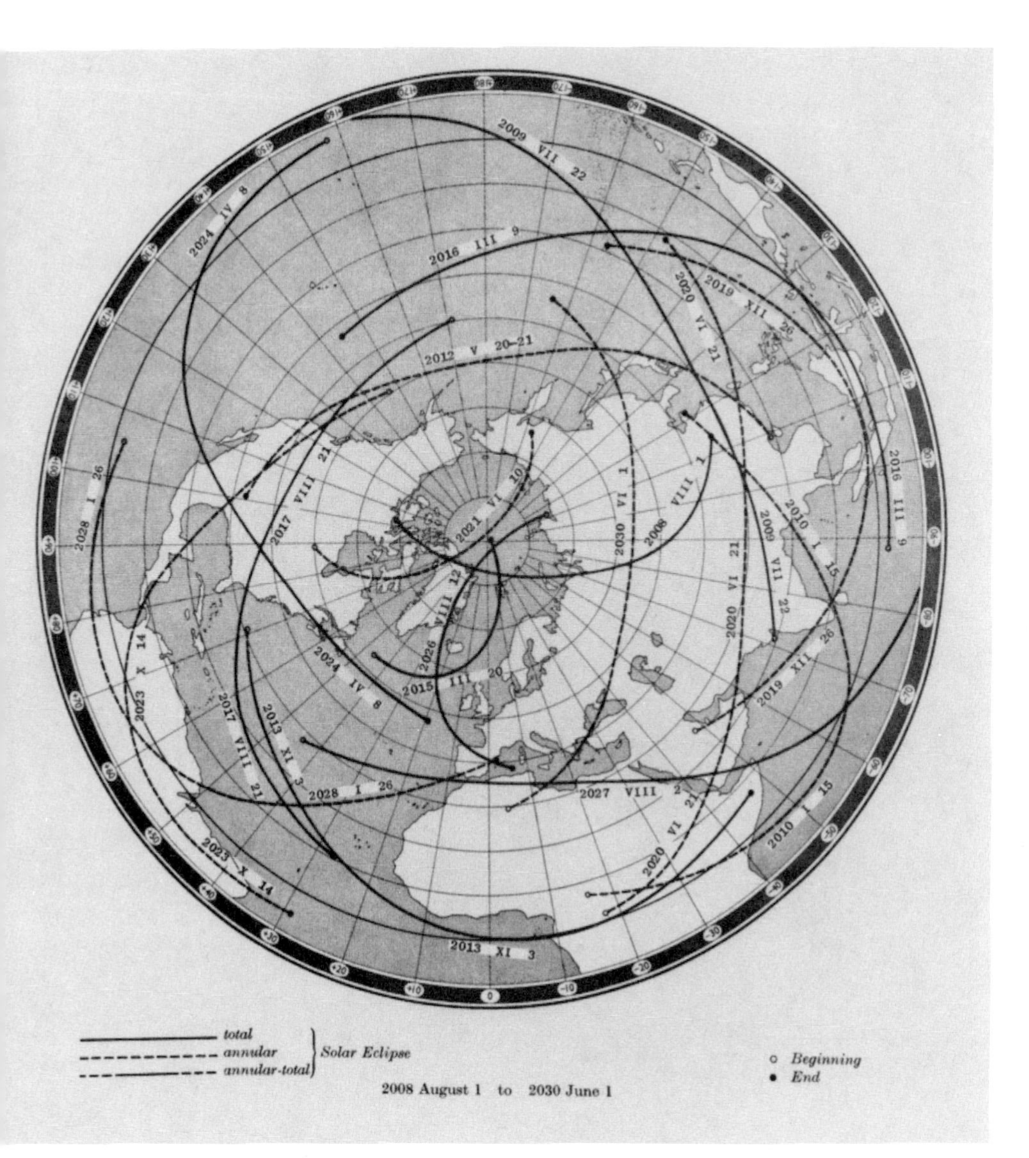

2008 August 1 to 2030 June 1

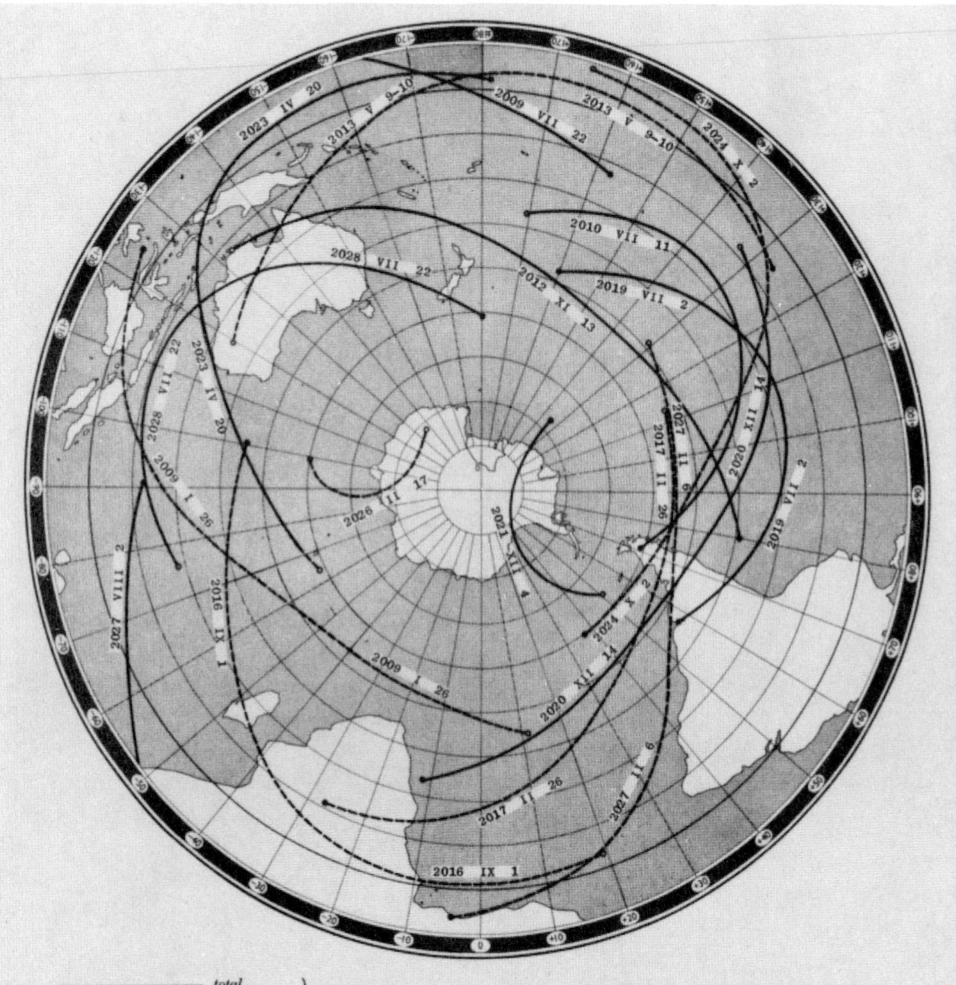

2009 January 26 to 2028 July 22

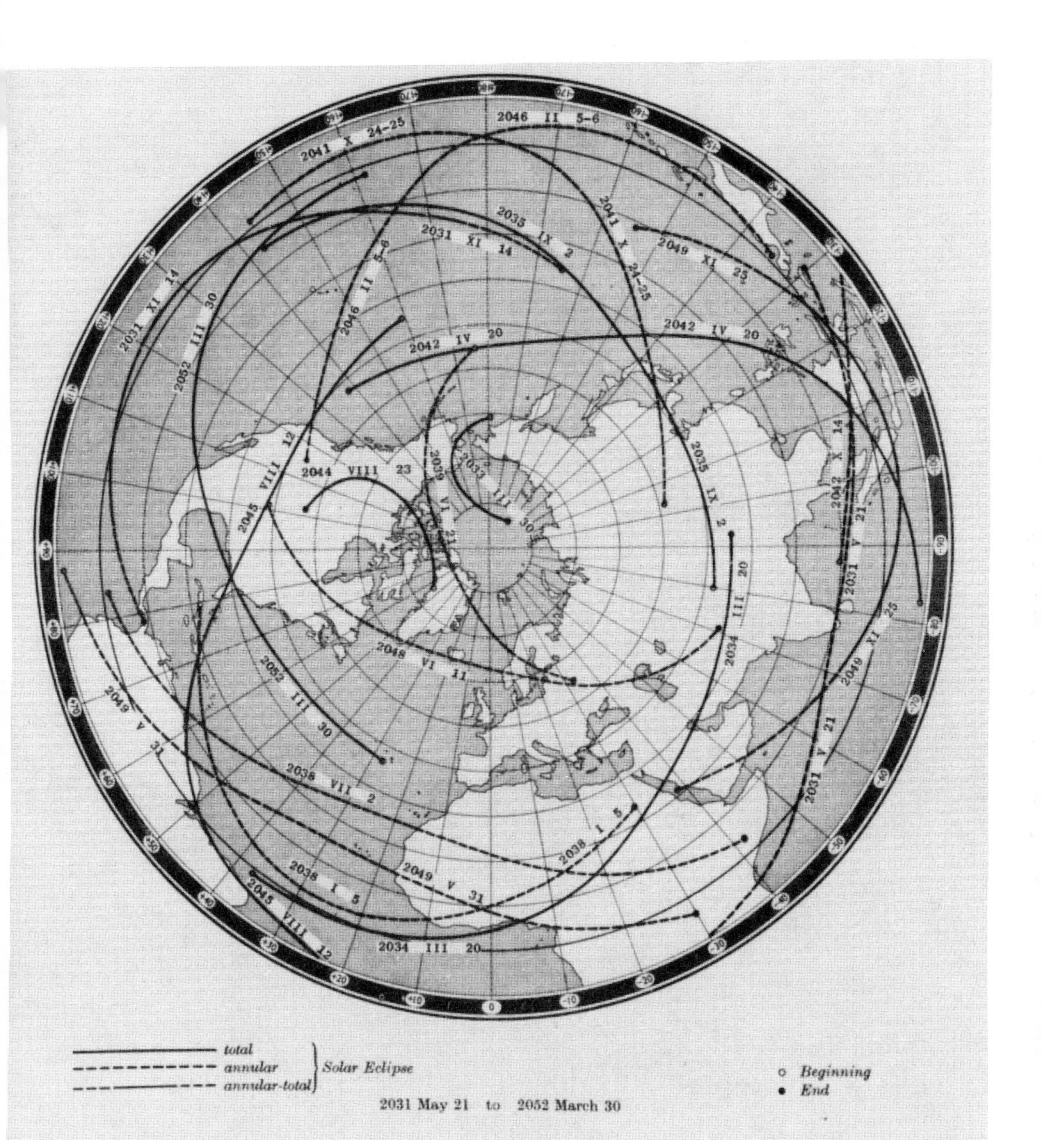

2031 May 21 to 2052 March 30

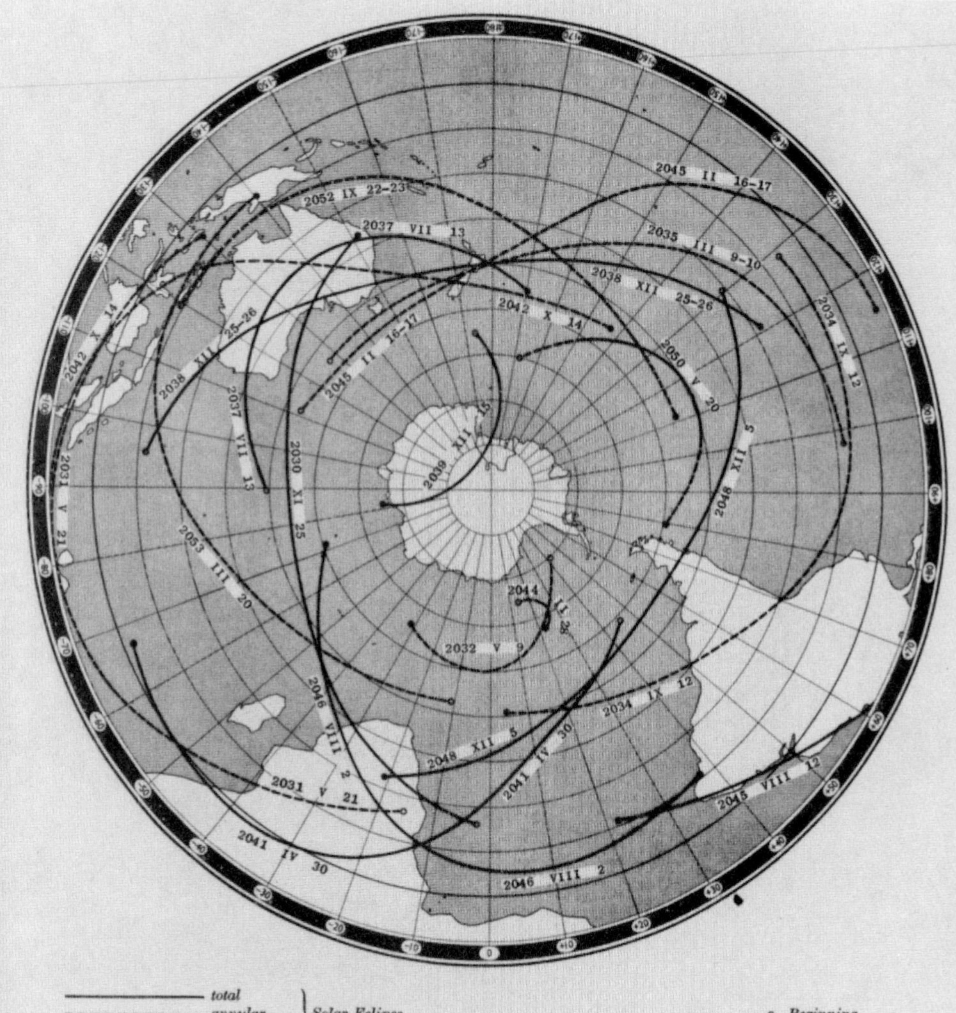

2030 November 25 to 2053 March 20

Appendix C

Chronology of Discoveries about the Sun

Year

Made during Solar Eclipses

Made at Other Times

2159–1948 B.C.

Legendary dates from China in the *Shu Ching* of the first recorded solar eclipse. In this myth, Chinese astronomers Hsi and Ho fail to prevent or predict or properly react to an eclipse and are sentenced by an angry emperor

1307 B.C.

First recorded observation of the corona (or prominences?) during a solar eclipse, inscribed on oracle bones in China: "three flames ate up the Sun, and a great star was visible"

1223 B.C., March 5

Oldest record of a verifiable solar eclipse, on a clay tablet found in the ruins of Ugarit (Syria)

1217 B.C.

Oldest Chinese record of a verifiable solar eclipse: inscriptions on an oracle bone dating from the Shang dynasty

585 B.C., May 28

A total eclipse in the midst of a battle between the Lydians and Medes scares both sides; hostilities are suspended, according to the Greek historian Herodotus (several other dates are possible)

sixth century B.C.

Babylonians (Chaldeans) are said to be able to predict eclipses of the Sun and Moon, supposedly based on cycles such as the saros, but more likely from the position of the Moon's nodes

450 B.C.?

Anaxagoras (Greece) is probably the first to realize that the Moon is illuminated by the Sun, thus providing a scientific explanation of eclipses

Philosopher Anaxagoras proposes that the Sun is a giant glowing ball of rock; he is banished from Athens for blasphemy

Year

Made during Solar Eclipses

Made at Other Times

431 B.C., August 3

Oldest European record of a verifiable solar eclipse (annular), by the Greek historian Thucydides

330 B.C.?

Aristotle (Greece) proposes that the Sun is a sphere of pure fire, unchanging and without imperfections

130 B.C.

Greek astronomer Hipparchus uses the position of the Moon's shadow during a solar eclipse to estimate the distance to the Moon (accurate to about 13%)

20 B.C.?

Liu Hsiang is first in China to explain that the Moon's motion hides the Sun to cause solar eclipses

150 A.D.?

Ptolemy (Alexandria) demonstrates the computation of solar and lunar eclipses based on their apparent motions rather than the periodic repetition of eclipses

334, July 17

Firmicus (Sicily) is first to report solar prominences, seen during an annular eclipse

418, July 19

First report of a comet discovered during a solar eclipse, seen by the historian Philostorgius in Asia Minor

ninth century

Shadow bands during a total eclipse are described for the first time, in the *Volospa*, part of the old German poetic edda

968, December 22

First clear description of the corona seen during a total eclipse, by a chronicler in Constantinople

1605

Johannes Kepler (Germany) is the first to comment scientifically on the solar corona, suggesting that it is light reflected from matter around the Sun (based on reports of eclipses; he never saw a total eclipse)

1609

Galileo Galilei (Italy) and Johann Fabricius (Holland) independently

Year

Made during Solar Eclipses

Made at Other Times

observe sunspots to be part of the Sun, disproving Aristotle's contention that the Sun is unchanging and free of "imperfections"

1683

Gian Domenico Cassini (Italy/France) explains zodiacal light as sunlight reflected from small solid particles in the plane of the solar system

1687

Isaac Newton (England) publishes his *Principia,* including the law of universal gravitation, which makes precise long-range eclipse prediction possible

1695

Edmond Halley (England) is first to notice that the reported times and places of ancient eclipses do not correlate with calculations backward from his era; he concludes correctly that the Moon's orbit has changed slightly (secular acceleration)

1706, May 12

An English ship captain named Stannyan, on vacation in Switzerland, reports a reddish streak (chromosphere? prominence?) along the rim of the Sun as the eclipse becomes total

1715, May 3

Edmond Halley (England), during an eclipse in England, is the first to report the phenomenon later known as Baily's Beads; also notes bright red prominences and the east-west asymmetry in the corona, which he attributes to an atmosphere on the Moon or Sun

1715

Gian Domenico Cassini (Italy/France) proposes that the light responsible for the corona also causes the zodiacal light

1724, May 22

Giacomo Filippo Maraldi (Italy/France) concludes that the corona is part of the Sun because the Moon traverses the corona during an eclipse

1733, May 13

Birger Wassenius (Sweden), observing an eclipse near Göteborg, is the first to report prominences visible to the unaided eye; he attributes them to the Moon

Year

Made during Solar Eclipses

Made at Other Times

1749

Richard Dunthorne (United Kingdom) calculates the approximate secular acceleration of the Moon based on Halley's 1693 findings

1774

Alexander Wilson (Scotland) shows that sunspots are depressions in the Sun's photosphere, not clouds above it or mountains or volcanic deposits on it

1795

William Herschel (United Kingdom) proposes that sunspots are holes in the Sun's hot clouds through which the dark, cool, solid surface of the Sun can be seen; he also suggests that this surface may be inhabited

1800

William Herschel (United Kingdom) founds the science of solar physics by measuring the temperature of various colors in the Sun's spectrum; he detects infrared radiation beyond the visible spectrum

1801

Johann Wilhelm Ritter (Germany), following the lead of Herschel, uses the spectrum of the Sun to establish the existence of ultraviolet radiation

1802

Dark (absorption) lines are discovered in the Sun's spectrum by William Wollaston (United Kingdom)

1806, June 16

José Joaquin de Ferrer (Spain), observing at Kinderhook, New York, gives the name *corona* to the glow of the faint outer atmosphere of the Sun seen during a total eclipse; he proposes that the corona must belong to the Sun, not the Moon, because of its great size

1817

Joseph Fraunhofer (Germany) independently discovers and catalogs many hundreds of dark lines in the Sun's spectrum

1820

Carl Wolfgang Benjamin Goldschmidt (Germany) calls attention to the shadow bands visible just before and after totality at some eclipses (based on the eclipse of November 19, 1816?)

Year

Made during Solar Eclipses

Made at Other Times

1824

Friedrich Wilhelm Bessel (Germany) introduces an easier way (Bessel functions) to make eclipse predictions

1833

David Brewster (United Kingdom) shows that some of the dark lines in the Sun's spectrum are due to absorption in the Earth's atmosphere, but most are intrinsic to the Sun

1836, May 15

Francis Baily (United Kingdom), during an annular eclipse in Scotland, calls attention to the brief bright beads of light that appear close to totality as the Sun's disk is blocked except for sunlight streaming through lunar valleys along the limb. This phenomenon becomes known as *Baily's Beads*

1837–1838

John F. W. Herschel (United Kingdom) and Claude Servais Mathias Pouillet (France) independently make the first quantitative measurements of the heat emitted by the Sun (about half the actual value). James David Forbes (United Kingdom) obtains a more accurate value in 1842, but it is considered less reliable

1842, July 8

Francis Baily (United Kingdom), at an eclipse in Italy, focuses attention on the corona and prominences and identifies them as part of the Sun's atmosphere

1843

Samuel Heinrich Schwabe (Germany), after a 17-year study, discovers the sunspot cycle of 10 years (now known to average 11 years)

1845

First clear photograph of the Sun: a daguerreotype by Hippolyte Fizeau and Léon Foucault (France)

1848

Julius Robert Mayer (Germany) shows through calculations that the Sun cannot shine a significant length of time by ordinary chemical burning; he incorrectly attributes the energy of the Sun to the heat released by the impact of meteoroids on the Sun

Year

Made during Solar Eclipses

Made at Other Times

1851, July 28

First astronomical photograph of a total eclipse: a daguerreotype by Berkowski at Königsberg, Prussia

1851, July 28

Robert Grant and William Swan (United Kingdom) and Karl Ludwig von Littrow (Austria) determine that prominences are part of the Sun because the Moon is seen to cover and uncover them as it moves in front of the Sun

1851, July 28

George B. Airy (United Kingdom) is the first to describe the Sun's chromosphere: he calls it the *sierra*, thinking that he is seeing mountains on the Sun, but he is actually seeing small prominences (spicules) that give the chromosphere a jagged appearance. Because of its reddish color, J. Norman Lockyer, in 1868, names this layer of the Sun's atmosphere the *chromosphere*

1852

Johann von Lamont (Germany), after 15 years of observations, discovers a 10.3-year activity cycle in the Earth's magnetic field, but does not correlate it with the sunspot cycle

1852

Edward Sabine (Ireland/United Kingdom), Rudolf Wolf (Switzerland), and Alfred Gautier (France) independently link the sunspot cycle to magnetic fluctuations on Earth; the study of solar-terrestrial relationships begins

1854

Hermann von Helmholtz (Germany) attributes the energy of the Sun to heat from gravitational contraction

1858

Richard C. Carrington (United Kingdom), studying sunspots, identifies the Sun's axis of rotation and discovers that the Sun's rotational period varies with latitude. He also discovers that the latitude of sunspots migrates toward the equator during the course of a sunspot cycle

1859, September 1

Richard C. Carrington and R. Hodgson (United Kingdom) are the first to observe a flare on the Sun. They both also note that a mag-

Year

Made during Solar Eclipses

Made at Other Times

netic storm in progress on Earth intensifies soon afterwards, but they refrain from connecting the two events

1859

Gustav Kirchhoff (Germany) uses spectroscopy to show that the surface of the Sun cannot be solid and that the Sun's atmosphere (which he identifies with the corona) is responsible for the dark lines in the Sun's spectrum

1860, July 18

First wet plate photographs of an eclipse; they require 1/30 of the exposure time of a daguerreotype

1860, July 18

Warren De La Rue (United Kingdom) and Angelo Secchi (Italy) use photography during a solar eclipse in Spain to demonstrate that prominences (and hence at least that region of the corona) are part of the Sun, not light scattered by the Earth's atmosphere or the edge of the Moon, because the corona looks the same from sites 250 miles apart

1859–1861

Gustav Kirchhoff and (in part) Robert Bunsen (Germany) identify 12 elements in the Sun based on lines in its spectrum

1861–1862

Gustav Kirchhoff (Germany) maps the solar spectrum

1868, August 18

During an eclipse seen from the Red Sea through India to Malaysia and New Guinea, prominences are first studied with spectroscopes and shown to be composed primarily of hydrogen by James Francis Tennant (United Kingdom), John Herschel (United Kingdom—son of John F. W. Herschel, grandson of William), Jules Janssen (France), Georges Rayet (France), and Norman Pogson (United Kingdom/India)

1868

Pierre Jules César Janssen (France) and J. Norman Lockyer (United Kingdom) independently demonstrate that prominences are part of the Sun (not the Moon) by observing them in days after the eclipse of August 18

1868

J. Norman Lockyer (United Kingdom) identifies a yellow spectral line in the Sun's corona as the signature of a chemical element as yet

Year

Made during Solar Eclipses

Made at Other Times

unknown on Earth. He later names it *helium*, after the Greek word *helios*, the Sun. Helium is first identified on Earth by William Ramsay in 1895

1869, August 7

Charles Augustus Young and William Harkness (United States) independently discover a new bright (emission) line in the spectrum of the Sun's corona, never before observed on Earth; they ascribe it to a new element and it is named *coronium*. In 1941, this green line is identified by Bengt Edlén (Sweden) as iron that has lost 13 electrons

1869

Anders Jonas Angström (Sweden) produces an improved map of the solar spectrum (using a diffraction grating rather than a prism) and introduces angstrom unit for measuring wavelength

1869

Thomas Andrews (Ireland) shows more conclusively than those before him that the Sun must be made essentially of hot gas

1870, December 2

Jules Janssen (France) uses a balloon to escape the German siege of Paris to study the December 22 eclipse in Algeria. He reaches Algeria, but the eclipse is clouded out

1870, December 22

Charles A. Young (United States), observing an eclipse in Spain, discovers that the chromosphere is the layer in the solar atmosphere that produces the dark lines in the Sun's spectrum

1870

Samuel P. Langley (United States) uses the Doppler Effect of the Sun's rotation to show that it displaces the dark (absorption) lines in the solar spectrum

1871, December 12

Jules Janssen (France) uses spectroscopy from an eclipse in India to propose that the corona consists of both hot gases and cooler particles and therefore is part of the Sun

1872, August 3

Charles A. Young (United States) observes a flare on the Sun with a spectroscope; he calls attention to its coincidence with a magnetic storm on Earth

Year

Made during Solar Eclipses

Made at Other Times

1874

Samuel P. Langley (United States) proposes that the bright granules in the Sun's photosphere are columns of hot gas rising from the interior and the dark interstices are cooler gases descending; he also proposes that the bright granules are responsible for almost all of the Sun's light

1871 and 1878

Jules Janssen (France) notices that the shape of the corona changes with the sunspot cycle. At sunspot maximum, the corona is rounder (1871); at sunspot minimum, the corona is more equatorial (1878). This discovery is the most convincing evidence that the corona is part of the Sun

1878, July 29

Height of search for intra-Mercurial planet Vulcan using eclipses to block the Sun. Several observers claim sightings, but they are never confirmed. The problem is finally resolved by Einstein in his general theory of relativity in 1916

1878, July 29

Samuel P. Langley and Cleveland Abbe (United States), observing from Pike's Peak in Colorado, and Simon Newcomb (United States), observing from Wyoming, notice coronal streamers extending more than 6 degrees from the Sun along the ecliptic and suggest that this glow is the origin of the zodiacal light

1882, May 17

A comet is discovered and photographed by Arthur Schuster (Germany/United Kingdom) during an eclipse in Egypt: first time a comet discovered in this way has been photographed

1884

Marie Alfred Cornu (France) applies Langley's work using the Doppler Effect of the Sun's rotation to distinguish between dark (absorption) lines created in the Sun's atmosphere and those created in the Earth's atmosphere

1887

Theodor von Oppolzer's (Austria) monumental *Canon of Eclipses* published, giving details of almost all solar and lunar eclipses from 1207 B.C. to A.D. 2161

Year

Made during Solar Eclipses

Made at Other Times

1887, August 19

Dmitri Ivanovich Mendeleev (Russia) uses a balloon to ascend above the cloud cover to an altitude of 11,500 feet (3.5 km) to observe an eclipse in Russia

1889

Henry Augustus Rowland (United States) produces an essentially modern map of the solar spectrum, identifying a total of 36 elements present in the Sun

1893 and 1894

George Ellery Hale (United States) and Henri Deslandres (France) independently develop spectroheliographs to photograph the Sun's chromosphere, prominences, and flares without waiting for eclipses

1894

William E. Wilson and P. L. Gray (Ireland) are the first to measure with reasonable accuracy the effective temperature of the Sun's photosphere: 11,200 °F (6,200 °C), about 800 °F (400 °C) too high

1899

Friedrich Karl Ginzel (Austria), Oppolzer's co-worker, uses data from the *Canon of Eclipses* for his *Special Canon of Solar and Lunar Eclipses*, which lists references in classical literature to eclipses between 900 B.C. and A.D. 600

1904

George Ellery Hale (United States) establishes on Mount Wilson in California the first large solar observatory

1908

George Ellery Hale (United States) shows that sunspots are regions with strong magnetic fields

1911

Albert Einstein (Germany), working on his general theory of relativity, proposes that gravity bends light and that this phenomenon might be observed during a solar eclipse

1916

Einstein publishes his complete general theory of relativity with a revised prediction for the gravitational deflection of starlight

Year

Made during Solar Eclipses

Made at Other Times

1919, May 29

Arthur S. Eddington (United Kingdom) and co-workers, observing a total solar eclipse from Principe and Brazil, confirm the bending of starlight by gravity as predicted by Einstein in his general theory of relativity

1922, September 21

William Wallace Campbell and Robert J. Trumpler (United States) reconfirm Einstein's relativistic bending of starlight during an eclipse in Wallal, Australia

1926

Arthur S. Eddington (United Kingdom) proposes that the Sun and stars derive their energy from nuclear reactions at their core

1930

Bernard Lyot (France) invents the coronagraph, which creates an artificial eclipse inside a telescope so that the corona can be studied outside of eclipses

1932, August 31

G. G. Cillié (United Kingdom) and Donald H. Menzel (United States) use eclipse spectra to show that the Sun's corona has a higher temperature (faster atomic motion) than the photosphere. Confirmed, with much higher temperatures, by R. O. Redman during an eclipse in South Africa on October 1, 1940

1951

Ludwig F. Biermann (Germany) discovers the solar wind, a stream of charged gas particles ejected from the Sun in all directions

1962–1972

NASA's Orbiting Solar Observatories (OSOs)

1972

Using data from an Orbiting Solar Observatory (OSO 4), Richard H. Munro and George L. Withbroe (United States) discover coronal holes, later shown by Werner M. Neupert and Victor Pizzo (United States) to be the source of solar wind because they correlate with magnetic storms on Earth

1973, June 30

Scientists use a Concorde supersonic passenger jet flying at 1,250 miles (2,000 km) an hour over Africa to extend the duration of solar eclipse

Year

Made during Solar Eclipses

Made at Other Times

totality to 74 minutes, 10 times longer than can be observed from the ground

1973–1974

Astronauts on NASA's Skylab orbiting laboratory study corona over a nine-month period using the Apollo Telescope Mount

1980

NASA's Solar Maximum Mission satellite is launched; proves that the Sun's energy output varies (craft, crippled by blown fuses after nine months, is repaired in orbit by Space Shuttle astronauts in 1984 and operates until 1989)

Notes

Chapter 1: The Experience of Totality

Epigraph: Donald H. Menzel and Jay M. Pasachoff, *A Field Guide to the Stars and Planets*, 2d ed., revised and enlarged (Boston: Houghton Mifflin, 1983), 409.

1. In sky observations, the western side of the Sun or Moon refers to the edge of the Sun or Moon closer to the western horizon. For observers in mid-northern latitudes, the Sun is usually to the south. When facing south, east is to the left and west is to the right. This south-looking orientation can briefly confuse readers who are used to maps that are oriented north, so that east is to the right and west to the left.
2. Special thanks to John R. Beattie of New York City, upon whose experience and description this chapter is based.

Chapter 2: The Great Celestial Cover-Up

Epigraph: Alfonso X, King of Castile as cited by Arthur Koestler, *The Sleepwalkers*, Universal Library (New York: Grosset & Dunlap, 1963), 69.

1. Alan D. Fiala, U.S. Naval Observatory, personal communication, April 1990.
2. The principal cause of the seasons is the tilt of the Earth's axis, not the Earth's rather modest change in distance from the Sun.
3. The Sun takes 365¼ days to appear to go through the constellations of the zodiac once around the sky, so some ancient peoples measured it or rounded it off to 360 days. Each day, the Sun goes one step around its great sky circle, which is why the Babylonians considered that a circle has 360 degrees. The Chinese considered that a circle had 365¼ degrees.
4. The use of the word *saros* to mean a 223-lunar-month eclipse cycle was erroneously introduced in 1691 by Edmond Halley when he applied it to the Babylonian eclipse cycle on the basis of a corrupt manuscript by the Roman naturalist Pliny. The Babylonian sign SAR has meaning as both a word and a number. As a word, it means (among other things) *universe.* As a number, it means *3,600,* signifying a large number. There is no evidence that the Babylonians ever applied *saros* to the 18-year eclipse cycle.
5. This period is called anomalistic because it derives from the anomaly, or irregularity, of lunar motion due to the Moon's elliptical orbit.
6. The reason for the change in latitude after 54 years 34 days is the 34-day difference from a calendar year, which changes the eclipse's position within the season, so the altitude of the Sun is significantly different and the shadow is cast farther north or south. The triple saros of 54 years 34 days

was known to the Chaldeans as well as the Greeks, who called it the *exeligmos*.

7. In a sense, it has not vanished altogether. With each period of 6,585.32 days, the Sun's position with respect to the node continues to slip westward without experiencing or causing any eclipses until, after about 5,500 years, it encounters the eclipse limit of the opposite node, and that saros may be said to be reborn. George van den Bergh, *Periodicity and Variation of Solar (and Lunar) Eclipses* (Haarlem: H. D. Tjeenk Willink, 1955).

Chapter 3: A Quest to Understand

Epigraph: Jack B. Zirker, *Total Eclipses of the Sun* (New York: Van Nostrand Reinhold, 1984), vi.

1. One of these 30 uprights is only half the diameter of all the others, as if to suggest 29½, the length of time in days that it takes the Moon to complete a cycle of phases.
2. There were also 56 chalk-filled Aubrey Holes just inside the embankment, but their function is still not generally agreed upon.
3. The hole for this stone was discovered in 1979 under the shoulder of the road that passes close to the Heel Stone. The original stone is gone. Michael W. Pitts, "Stones, Pits and Stonehenge," *Nature* 290 (March 5, 1981): 46–47.
4. The limits of the Moon's motion north and south of the ecliptic can differ by as much as 10 minutes of arc from the mean inclination of 5 degrees 8 minutes because of gravitational perturbations on the Moon caused by the Sun and the equatorial bulge of the Earth.
5. How Stonehenge may have been used as a computer to predict eclipses is proposed and explained in Gerald S. Hawkins (with John B. White), *Stonehenge Decoded* (Garden City, N.Y.: Doubleday, 1965); Gerald S. Hawkins, *Beyond Stonehenge* (New York: Harper & Row, 1973); and Fred Hoyle, *On Stonehenge* (San Francisco: W. H. Freeman, 1977).
6. James Legge, ed. and trans., *The Chinese Classics*, vol. 3, *The Shoo King* [Shu Ching] (Hong Kong: Hong Kong University Press, 1960), pt. 1, ch. 2, para. 3–8 (pp. 18–22). Legge romanizes Hsi's name as He. In some recountings, Hsi's name appears as Hi.
7. Legge, *The Shoo King*, pt. 3, bk. 4, ch. 2, para. 4 (pp. 165–166).
8. Joseph Needham and Wang Ling, *Science and Civilisation in China*, vol. 3, *Mathematics and the Sciences of the Heavens and the Earth* (Cambridge: Cambridge University Press, 1959), 422. For the record, there were no total solar eclipses visible in China through this four-year period, and no partials of any consequence. Perhaps, if there is any historical foundation to this story, the court astronomer predicted that a major eclipse would occur and it did not, or perhaps the problem concerned lunar eclipses.
9. Needham and Ling, *Mathematics and the Sciences of the Heavens and the Earth*, 409. Most of this information is also available in Colin A. Ronan's abridgment of Needham's work, *The Shorter Science and Civilisation in China*, vol. 2 (Cambridge: Cambridge University Press, 1981).

10. Anthony F. Aveni, *Skywatchers of Ancient Mexico* (Austin: University of Texas Press, 1980), 181.
11. Bernardino de Sahagún, *Florentine Codex: General History of the Things of New Spain*, bk. 7, *The Sun, Moon, and Stars, and the Binding of the Years*, trans. Arthur J. O. Anderson and Charles E. Dibble (Santa Fe, N.M.: School of American Research; Salt Lake City: University of Utah, 1953), 36, 38.

Chapter 4: Eclipses in Mythology

Epigraph: Francis Baily, "Some Remarks on the Total Eclipse of the Sun, on July 8th, 1842," *Memoirs of the Royal Astronomical Society*, vol. 15 (1846), 6.

1. Viktor Stegemann, "Finsternisse," in *Handwörterbuch des Deutschen Aberglaubens*, Bd. 2, Hanns Bächtold-Stäubli, ed., (Berlin: W. de Gruyter, 1930), columns 1509–1526. Germanic eclipse lore described in this chapter comes from this article, unless otherwise noted.
2. Arthur Berriedale Keith, *Indian Mythology*, vol. 6 of *The Mythology of All Races* (Boston: Marshall Jones, 1917), 151, 192.
3. Wilhelm Max Müller, *Egyptian Mythology*, vol. 12 of *The Mythology of All Races* (Boston: Marshall Jones, 1918), 33, 90, 124–125.
4. Paul Yves Sébillot, *Le folk-lore de France*, t. 1, *Le ciel et la terre* (Paris: Librairie orientale & américaine, 1904), 40.
5. John C. Ferguson, *Chinese Mythology*, vol. 8 of *The Mythology of All Races* (Boston: Marshall Jones, 1928), 84. Hartley Burr Alexander, *Latin-American Mythology*, vol. 11 of *The Mythology of All Races* (Boston: Marshall Jones, 1920), 319. Mardiros H. Ananikian, *Armenian Mythology*, vol. 7 of *The Mythology of All Races* (Boston: Marshall Jones, 1925), 48.
6. Keith, *Indian Mythology*, 232–233.
7. Ananikian, *Armenian Mythology*, 48.
8. Keith, *Indian Mythology*, 234.
9. Joseph Needham and Wang Ling, *Science and Civilisation in China*, vol. 3, *Mathematics and the Sciences of the Heavens and the Earth* (Cambridge: Cambridge University Press, 1959), 228.
10. Hartley Burr Alexander, *North American Mythology*, vol. 10 of *The Mythology of All Races* (Boston: Marshall Jones, 1916), 25, 277.
11. James George Frazer, *Balder the Beautiful*, vol. 1, The Golden Bough, vol. 10 (London: Macmillan, 1930), 162.
12. Mabel Loomis Todd, *Total Eclipses of the Sun*, rev. ed. (Boston: Little, Brown, 1900), 131.
13. Alexander, *North American Mythology*, 255.
14. Sébillot, *Le folk-lore de France*, 52. The chronicler was Jean Juvénal des Ursins.
15. Keith, *Indian Mythology*, 234.
16. A free translation from the "Second Soir." Compare Fontenelle, *A Plurality of Worlds*, John Glanvill, trans. (England: Nonesuch Press, 1929), with the excerpt in François Arago, *Popular Astronomy*, vol. 2, translated by

W. H. Smyth and Robert Grant (London: Longman, Brown, Green, Longmans, and Roberts, 1858), 349.

17. Alexander, *Latin-American Mythology*, 277–278.
18. James George Frazer, *The Magic Art*, vol. 1, The Golden Bough, vol. 1 (London: Macmillan, 1926), 311.
19. Patrick Menget, "30 juin 1973: station de Surinam," *Soleil est mort: l'éclipse totale de soleil du 30 juin 1973* (Nanterre, France: Laboratoire d'ethnologie et de sociologie comparative, 1979), 119–142.
20. Alexander, *Latin-American Mythology*, 135.
21. Alexander, *Latin-American Mythology*, 82.

Chapter 5: Strange Behavior of Man and Beast

Epigraph: John Milton, *Paradise Lost*, bk. 1, lines 594 and 597–599. See John Milton, *The Complete Poems* (New York: Crown Publishers, 1936), 24.

1. Herodotus, *The History*, vol. 1, George Rawlinson, trans. Everyman's Library, vol. 405 (London: J. M. Dent, 1910), bk. 1, ch. 74 (pp. 36–37).
2. Robert R. Newton lists three annular eclipses seen in the region during a 50-year period which he feels are equally likely to have given rise to this story, although an annular eclipse is not nearly as spectacular as one that is total. *Ancient Astronomical Observations and the Accelerations of the Earth and Moon* (Baltimore: Johns Hopkins Press, 1970), 94–97.
3. Herodotus, *The History*, vol. 2, George Rawlinson, trans. Everyman's Library, vol. 406 (London: J. M. Dent, 1910), bk. 7, ch. 37 (p. 136).
4. Plutarch, *The Rise and Fall of Athens: Nine Greek Lives*, Ian Scott-Kilvert, trans. (Baltimore: Penguin Books, 1960), 201–202. Ironically, Pericles' raid was disastrous for the Athenian forces; they fell victim to the plague. Pericles was fined and temporarily stripped of power.
5. Thucydides, *History of the Peloponnesian War*, Richard Crawley, trans. (New York: E. P. Dutton, 1910), bk. 2, para. 28. Stars would not have been visible at Athens. Perhaps Thucydides heard reports from where the eclipse was annular and incorporated them into his account.
6. From the chapter "Second Soir" in Bernard Le Bovier de Fontenelle, *Entretiens sur la pluralité des mondes* (Paris: Chez la veuve C. Blageart, 1686), cited in François Arago, *Popular Astronomy*, vol. 2, translated by W. H. Smyth and Robert Grant (London: Longman, Brown, Green, Longmans, and Roberts, 1858), 359–360.
7. Mabel Loomis Todd, *Total Eclipses of the Sun*, rev. ed. (Boston: Little, Brown, 1900), 141–142.
8. F. Richard Stephenson and David H. Clark, *Applications of Early Astronomical Records*, Monographs on Astronomical Subjects, no. 4 (New York: Oxford University Press, 1978), 9.
9. Stephenson and Clark, *Applications*, 14.
10. François Arago, *Popular Astronomy*, 359.
11. Arago, *Popular Astronomy*, 362.
12. William J. S. Lockyer, "The Total Eclipse of the Sun, April 1911, as

Observed at Vavau, Tonga Islands," *Astronomy*, vol. 2, Bernard Lovell, ed. The Royal Institution Library of Science (Barking, Essex: Elsevier Publishing, 1970), 190–191.

Chapter 6: Anatomy of the Sun

Epigraph: Amos 8:9, *New American Standard Bible.*

1. Most of the statistics in this chapter are from Robert W. Noyes, *The Sun, Our Star* (Cambridge, Mass.: Harvard University Press, 1982).
2. To generate and sustain a hydrogen-to-helium fusion reaction, the core of a star must have a temperature of at least 18 million °F (10 million °C). To have enough gravity to generate sufficient pressure to obtain this high a core temperature, a star must have at least 8 percent the mass of the Sun.
3. The expressions "fire-ocean" to describe the chromosphere and "flame-brushes" to describe features in the corona were used by Agnes M. Clerke, *A Popular History of Astronomy during the Nineteenth Century*, 4th ed. (London: A. and C. Black, 1902), 68, 175.

Chapter 7: Lessons from Eclipses

Epigraph: Francis Baily, "Some Remarks on the Total Eclipse of the Sun, on July 8th, 1842," *Memoirs of the Royal Astronomical Society*, vol. 15 (1846), 4.

1. Francis Baily, "On a Remarkable Phenomenon That Occurs in Total and Annular Eclipses of the Sun," *Memoirs of the Royal Astronomical Society* 10 (1838): 1–42. Baily, searching back through the records, realized that Edmond Halley in 1715 and many other observers had seen and reported this beadlike apparition before him. Among the previous observers of the beads, he named José Joaquin de Ferrer, who also described the corona and first called it by that name.
2. This and the following Baily quotes are from Francis Baily, "Some Remarks on the Total Eclipse of the Sun, on July 8th, 1842. Pp. 1–8.
3. Agnes M. Clerke, "Baily, Francis," *The Dictionary of National Biography*, vol. 1 (London: Oxford University Press, 1921), 903.
4. Joseph Needham and Wang Ling, *Science and Civilisation in China*, vol. 3, *Mathematics and the Sciences of the Heavens and the Earth* (Cambridge: Cambridge University Press, 1959), 423.
5. José Joaquin de Ferrer, "Observations of the Eclipse of the Sun, June 16th, 1806, Made at Kinderhook, in the State of New-York," *Transactions of the American Philosophical Society* 6 (1809): 264–275.
6. Dorrit Hoffleit, *Some Firsts in Astronomical Photography* (Cambridge, Mass.: Harvard College Observatory, 1950). The first successful photograph of the uneclipsed Sun, also a daguerreotype, was achieved by the French scientists Hippolyte Fizeau and Léon Foucault in 1845.
7. De La Rue made that discovery in 1861. Earlier evidence for sunspots as depressions had come from observations by Scottish astronomer Alexan-

der Wilson in 1774. He noted that the geometry of sunspots seemed to change as they were seen from different angles as the Sun's rotation carried them across the solar disk and toward the limb.

8. The Sun as a sphere of hot gas had been proposed independently by Angelo Secchi and John Frederick William Herschel (William's son) in 1864.
9. J. Norman Lockyer, "On Recent Discoveries in Solar Physics Made by Means of the Spectroscope," *Astronomy*, vol. 1, Bernard Lovell, ed. The Royal Institution Library of Science (Barking, Essex: Elsevier Publishing, 1970), 90.
10. Alfred Fowler, "Sir Norman Lockyer, K.C.B., 1836–1920," *Proceedings of the Royal Society of London*, ser. A, 104 (Dec. 1, 1923): i–xiv.
11. Auguste Comte, *The Essential Comte, Selected from Cours de philosophie positive*, Margaret Clarke, trans. (London: Croom Helm, 1974), 74, 76.
12. A. J. Meadows, *Science and Controversy: A Biography of Sir Norman Lockyer* (London: Macmillan, 1972), 53.
13. Lockyer, "On Recent Discoveries," 101–102.
14. "A gravely mutilated state" is the picturesque description found in *The Flammarion Book of Astronomy*, Gabrielle Camille Flammarion and André Danjon, eds., Annabel and Bernard Pagel, trans. (New York: Simon and Schuster, 1964), 227.

Chapter 8: A Magic Shadow Show

Epigraph: Arthur S. Eddington as cited in Allie Vibert Douglas, *The Life of Arthur Stanley Eddington* (London: T. Nelson, 1956), 44.

1. The angular displacement of a star is inversely proportional to the angular distance of that star from the Sun's center.
2. In 1911, Einstein had asked Freundlich to investigate another aspect of the emerging general theory of relativity: the motion of Mercury. Freundlich reviewed the anomaly in Mercury's motion, recognized since the work of Le Verrier in 1859, and reconfirmed that Mercury was indeed deviating slightly from the law of gravity as formulated by Newton. Freundlich's results, which matched Einstein's relativistic recalculation for the precession of Mercury's orbit, were published in 1913 over the objections of his superiors.
3. To make certain that the telescope optics and the camera system introduced no unknown deflection in the positions of stars, it was important to have for comparison with the eclipse plate a picture of that same region taken with the same equipment when the Sun was not present. No such plates were available.
4. Charles Dillon Perrine, an American who was directing the Argentine National Observatory, prepared to test the bending of starlight during the October 10, 1912 eclipse in Brazil, but was rained out.
5. Ronald W. Clark, *Einstein, the Life and Times* (London: Hodder and Stroughton, 1973), 176.

6. Freundlich's team was not alone in the Crimea. An American team from the Lick Observatory also journeyed to Russia to test relativity, but they were clouded out. In 1918, Freundlich left the Royal Observatory to work full time with Einstein at the Kaiser Wilhelm Institute. In 1920, he was appointed observer and then chief observer and professor of astrophysics at the newly created Einstein Institute at the Astrophysical Observatory, Potsdam. Freundlich continued to be plagued by miserable luck on his eclipse expeditions to measure the deflection of starlight. He returned empty-handed in 1922 and 1926 because of bad weather. He finally got to see an eclipse in Sumatra in 1929, although he obtained a deflection (now known to be erroneous) considerably greater than Einstein predicted. When Hitler came to power, Freundlich left Germany and eventually settled in Scotland, where he changed his name to Finlay-Freundlich, based on his mother's maiden name, Finlayson. He built the first Schmidt-Cassegrain telescope, the prototype for almost all large photographic survey telescopes today.
7. Banesh Hoffmann with the collaboration of Helen Dukas, *Albert Einstein, Creator and Rebel* (New York: Viking Press, 1972), 116.
8. Hoffmann, *Albert Einstein*, 125.
9. Robert V. Pound and Glen A. Rebka, Jr., "Resonant Absorption of the 14.4-kev Gamma Ray from 0.10-microsecond Fe^{57}," *Physical Review Letters* 3 (December 15, 1959): 554–556. The gravitational redshift was detected previously, based on Einstein's suggestion, in light emitted from extremely dense white dwarf stars, but to nowhere near the same degree of certainty.
10. The Lick team was working with mediocre equipment and improvised mounts. Their excellent regular equipment had stood ready under cloudy skies in Russia in 1914, but it was too cumbersome to transport home after the outbreak of World War I. After the war, the shipment home was delayed, and the equipment was not available for the eclipse of 1918.
11. The "Newtonian prediction" is calculated on the basis that light energy has a mass equivalent (Einstein's $E = mc^2$). That mass is then treated as ordinary matter using Newton's equations for gravity.
12. Allie Vibert Douglas, *The Life of Arthur Stanley Eddington* (London: T. Nelson, 1956), 40–41.
13. Clark, *Einstein*, 226.
14. Frank W. Dyson, Andrew C. D. Crommelin, and Arthur S. Eddington, "Joint Eclipse Meeting of the Royal Society and the Royal Astronomical Society," *The Observatory* 42 (Nov. 1919): 389–398. The article includes dissenting comments from scientists in the audience. The article based on eclipse results was Frank W. Dyson, Arthur S. Eddington, and Charles R. Davidson, "A Determination of the Deflection of Light by the Sun's Gravitational Field, From Observations Made at the Total Eclipse of May 29, 1919," *Philosophical Transactions of the Royal Society of London*, ser. A, 220 (1920): 291–333. A summary of the eclipse results had appeared the week after the joint meeting: Andrew C. D. Crommelin, "Results of the Total Solar Eclipse of May 29 and the Relativity Theory," *Nature* 104 (Nov.

1. Isabel Martin Lewis, *A Handbook of Solar Eclipses* (New York: Duffield, 1924), 98.

Chapter 12: Eclipse Photography

Epigraph: Stephen J. Edberg, personal communication, March 1990.

1. George B. Airy, "On the Total Solar Eclipse of 1851, July 28," *Astronomy*, vol. 1, Bernard Lovell, ed. The Royal Institution Library of Science (Barking, Essex: Elsevier Publishing, 1970), 4.
2. Dennis di Cicco, personal communication, July 1990.
3. Michael Covington, *Astrophotography for the Amateur* (Cambridge: Cambridge University Press, 1988).
4. Astronomy software for home computers that provides this information:

Program	*Supplier*	*Computer*
Sky Travel	Deltron Ltd. 155 Deer Hill Road Lebanon, NJ 08833 201-236-2098	Atari ST Apple II Commodore Macintosh
Voyager	Carina Software 830 Williams Street San Leandro, CA 94577 415-352-7328	Macintosh
The Observatory	Lightspeed Software 2124 Kittredge Street Suite 185 Berkeley, CA 94704 415-486-1165	Apple II
LodeStar Plus	Zephyr Services 1900 Murray Avenue Department A Pittsburgh, PA 15217 800-533-6666	IBM
SuperStar	Astronomical Society of the Pacific 390 Ashton Avenue San Francisco, CA 94112 415-661-8660	IBM

Chapter 13: The Pedigree of an Eclipse

Epigraph: George B. Airy, "On the Total Solar Eclipse of 1851, July 28," *Astronomy*, vol. 1, Bernard Lovell, ed. The Royal Institution Library of Science (Barking, Essex: Elsevier Publishing, 1970), 1.

Chapter 14: The Eclipse of July 11, 1991

Epigraph: Isabel Martin Lewis, *A Handbook of Solar Eclipses* (New York: Duffield, 1924), 3.

1. At present, the leading (but not the only) theory of the Moon's formation involves a collision between a large planetesimal and the Earth.
2. Mauna Kea rises 19,680 feet (6,000 m) from the floor of the Pacific Ocean to the surface of the sea and then continues another 13,756 feet (4,192 m) into the sky.
3. In the case of the total eclipse of July 11, 1991, the axis of the shadow cone passes just 17.4 miles (28 km) south of the center of the Earth. Joe Rao, *Your Guide to the Great Solar Eclipse of 1991* (Cambridge, Mass.: Sky Publishing, 1989), 58.
4. Fred Espenak calculates that the shadow of the July 11, 1991, total eclipse covers 0.69 percent of the Earth's surface. "Predictions for the Total Solar Eclipse of 1991," *Journal of the Royal Astronomical Society of Canada* 83 (June 1989): 162. The number of people along the path of totality of this eclipse was estimated by Rao, p. 14.

Glossary

annular eclipse A central eclipse of the Sun in which the angular diameter of the Moon is too small to completely cover the disk of the Sun and a thin ring *(annulus)* of the Sun's bright apparent surface surrounds the dark disk of the Moon. An annular eclipse is actually a kind of partial eclipse of the Sun. There are more annular eclipses than total eclipses.

annular-total eclipse A solar eclipse that begins as an annular eclipse, changes to a total eclipse along its path, and then returns to annular before the end of the eclipse path.

anomalistic month The time it takes (27.55 days) for the Moon to orbit the Earth as measured from its closest point to Earth (perigee) to its farthest point (apogee) and back to its closest point again.

anomalistic year The time it takes (365.26 days) for the Earth to orbit the Sun as measured from its closest point to the Sun (perihelion) to its farthest point (aphelion) and back to its closest point again.

aphelion The point for any object orbiting the Sun where it is farthest from the Sun.

apogee The point for any object orbiting the Earth where it is farthest from the Earth.

arc minute An angular measurement: 1 minute of arc is 60 seconds of arc and 1/60 of a degree of arc.

arc second An angular measurement: 1 second of arc is 1/60 of a minute of arc and 1/3600 of a degree of arc.

ascending node (of the Moon) The point on the Moon's orbit where it crosses the ecliptic (orbit of the Earth) going north.

Baily's Beads An effect seen just before and after the total phase of a solar eclipse in which the Moon hides all the light from the Sun's disk except for a few bright points of sunlight passing through valleys at the rim of the Moon.

central eclipse (of the Sun) An eclipse in which the axis of the Moon's shadow touches the Earth. A central eclipse can be either total or annular.

chromosphere The reddish lower atmosphere of the Sun just above the photosphere. The chromosphere is only about 1,600 miles (2,500 km) thick. The temperature at the base of the chromosphere is about 7,200 °F (4,000 °C).

contact (in a solar eclipse) Special numbered stages of a solar eclipse. In a total (or annular) eclipse, **first contact** occurs when the leading edge of the

Moon appears tangent to the western rim of the Sun, initiating the eclipse; **second contact** occurs when the Moon's leading edge appears tangent to the eastern rim of the Sun, initiating the total (or annular) phase of the eclipse; **third contact** occurs when the Moon's trailing edge appears tangent to the western rim of the Sun, concluding the total (or annular) phase of the eclipse; and **fourth contact** occurs when the trailing edge of the Moon appears tangent to the eastern rim of the Sun, concluding the eclipse. In a partial eclipse, there are only first and fourth contacts. These same contact points are used to describe lunar eclipses, planet transits across the face of the Sun, satellite transits across the face of a planet, and binary star transits across the face of one another.

core (of the Sun) The central region of the Sun where it produces its energy in a nuclear fusion reaction by converting hydrogen into helium at a temperature of 27 million °F (15 million °C).

corona The rarefied upper atmosphere of the Sun that appears as a white halo around the totally eclipsed Sun. Speeds of atomic particles in the corona give it a temperature sometimes exceeding 2 million °F (1.1 million °C).

coronagraph A special telescope that produces an artificial solar eclipse by masking the Sun's apparent surface with an opaque disk; invented by Bernard Lyot in 1930.

degree of obscuration The fraction of the area of the Sun's disk obscured by the Moon at eclipse maximum, usually expressed as a percentage. (Degree of obscuration is not the same as the magnitude of an eclipse, which is the fraction of the Sun's diameter that is covered).

descending node (of the Moon) The point on the Moon's orbit where it crosses the ecliptic (orbit of the Earth) going south.

draconic month The time it takes (27.21 days) for the Moon to orbit the Earth as measured from ascending node through descending node and back to ascending node again. Eclipses of the Sun and Moon can only take place near a node.

eclipse limit (for the Sun) The maximum angular distance that the Sun can be from a node of the Moon and still be involved in an eclipse seen from Earth. For partial eclipses, the limit ranges from 15°21′ to 18°31′ according to the varying angular sizes of the Moon and Sun due to the elliptical orbits of the Moon and Earth. For a central eclipse, the maximum and minimum limits are 11°50′ and 9°55′.

eclipse season The period of time in which the apparent motion of the Sun places it close enough to a node of the Moon so that an eclipse is possible. The Sun crosses the ascending and descending nodes of the Moon in a period of 346.62 days, so eclipse seasons occur about 173.3 days apart. Depending on where the Sun is on the ecliptic (how fast the Earth is moving), a solar eclipse season may last from 31 to 37 days.

eclipse year The time it takes (346.62 days) for the apparent motion of the Sun to carry it from ascending node to descending node and back to ascending node of the Moon.

ecliptic The apparent annual path of the Sun around the star field as seen from the Earth as the Earth orbits the Sun in the course of a year. (Thus the ecliptic is the plane of the Earth's orbit around the Sun.) The Sun's apparent path is called the ecliptic because all eclipses of the Sun and Moon occur on or very close to this track in the sky.

exeligmos An eclipse repetition cycle of 54 years 34 days, equal to three saros cycles and often called the triple saros. After one exeligmos cycle, a solar eclipse returns to almost the same longitude, but occurs about 600 miles (1,000 km) north or south of its predecessor.

flare Intense brightening in the upper atmosphere of the Sun that erupts vast amounts of charged particles into space. Flares can reach temperatures of 36 million °F (20 million °C).

inex A period of 10,571.95 days (29 years less 20.1 days), after which another eclipse of the Sun or Moon will occur (although not of the same type, such as total). This period equals 358 synodic months and 388.5 draconic months.

librations (of the Moon) Factors which, over time, allow Earth-based observers to see more than half of the Moon's surface. Because the Moon rotates and revolves in the same period of time, it always exposes the same face to Earth. As time passes, however, observers on Earth view 59 percent rather than 50 percent of the Moon's surface, seeing around the Moon's eastern and western limbs and over the poles, principally because of the Moon's elliptical orbit, the observer's position on Earth, and the Moon's position above or below the ecliptic.

lunation The time it takes (29.53 days) for the Moon to complete a phasing cycle (also called a synodic period).

magnitude (of a solar eclipse) The fraction of the apparent diameter of the solar disk covered by the Moon at eclipse maximum. Eclipse magnitude is usually expressed as a decimal fraction: below 1.000 is a partial eclipse; 1.000 or above is a total eclipse. (The magnitude of a solar eclipse is not the same as the degree of obscuration, which is the percentage of the area of the Sun's disk that is covered.)

maximum eclipse The moment in a solar eclipse when the shadow of the Moon passes closest to the center of the Earth. This is also the instant when the greatest fraction of the Sun's disk is obscured.

mid-eclipse The instant in a central solar eclipse halfway between second and third contacts.

New Moon The phase of the Moon when it is most nearly in conjunction with the Sun (also called dark-of-the-Moon). Solar eclipses can occur only at New Moon. (In ancient times, New Moon had a different meaning: the crescent Moon when it became visible after dark-of-the-Moon.)

Selected Bibliography

Allen, David, and Carol Allen. *Eclipse*. Sydney: Allen & Unwin, 1987.

Arago, François. *Popular Astronomy*. Translated by W. H. Smyth and Robert Grant. 2 vols. London: Longman, Brown, Green, Longmans, and Roberts, 1858.

Ashbrook, Joseph. *The Astronomical Scrapbook*. Edited by Leif J. Robinson. Cambridge: Cambridge University Press; Cambridge, Mass.: Sky Publishing, 1984.

Astrophotography Basics. Kodak Publication no. P-150. Rochester, N.Y.: Eastman Kodak, 1988.

Aveni, Anthony F. *Skywatchers of Ancient Mexico*. Austin: University of Texas Press, 1980.

Baily, Francis. "On a Remarkable Phenomenon that Occurs in Total and Annular Eclipses of the Sun." *Memoirs of the Royal Astronomical Society* 10 (1838): 1–40.

———. "Some Remarks on the Total Eclipse of the Sun, on July 8th, 1842." *Memoirs of the Royal Astronomical Society* 15 (1846): 1–8.

Brewer, Bryan. *Eclipse*. Seattle: Earth View, 1978.

Chambers, George F. *The Story of Eclipses*. Library of Valuable Knowledge. New York: D. Appleton, 1912.

Clerke, Agnes M. *A Popular History of Astronomy during the Nineteenth Century*. 4th ed. London: A. and C. Black, 1902.

Couderc, Paul. *Les éclipses*. Que sais-je? no. 940. Paris: Presses universitaires de France, 1961.

Covington, Michael. *Astrophotography for the Amateur*. Cambridge: Cambridge University Press, 1988.

Douglas, Allie Vibert. *The Life of Arthur Stanley Eddington*. London: T. Nelson, 1956.

Dyson, Frank, and Richard v.d.R. Woolley. *Eclipses of the Sun and Moon*. Oxford: Clarendon Press, 1937.

Espenak, Fred. *Fifty Year Canon of Solar Eclipses: 1986–2035*. Belmont, Mass.: Sky Publishing, 1987. NASA Reference Publication 1178 rev.

———. "Predictions for the Total Solar Eclipse of 1991." *Journal of the Royal Astronomical Society of Canada* 83 (June 1989): 157–178.

Fiala, Alan D., James A. DeYoung, and Marie R. Lukac. *Solar Eclipses, 1991–2000*. U.S. Naval Observatory Circular no. 170. Washington, D.C.: U.S. Naval Observatory, 1986.

Flammarion, Camille. *The Flammarion Book of Astronomy*. Edited by Gabrielle Camille Flammarion and André Danjon. Translated by Annabel and Bernard Pagel. New York: Simon and Schuster, 1964.

Francillon, Gérard, and Patrick Menget, eds. *Soleil est mort: l'éclipse totale de soleil du 30 juin 1973*. Nanterre: Laboratoire d'ethnologie et de sociologie comparative (Récherches thématiques 1), 1979.

Johnson, Samuel J. *Eclipses, Past and Future; with General Hints for Observing the Heavens*. Oxford: J. Parker, 1874.

Joslin, Rebecca R. *Chasing Eclipses: The Total Solar Eclipses of 1905, 1914, 1925*. Boston: Walton Advertising and Printing, 1929.

Kudlek, Manfred, and Erich H. Mickler. *Solar and Lunar Eclipses of the Ancient Near East from 3000 B.C. to 0 with Maps*. Neukirchen-Vluyn, Germany: Butzon & Bercker Kevelaer, 1971.

Lang, Kenneth R., and Owen Gingerich, eds. *A Source Book in Astronomy and Astrophysics, 1900–1975*. Cambridge, Mass.: Harvard University Press, 1979.

Lewis, Isabel Martin. *A Handbook of Solar Eclipses*. New York: Duffield, 1924.

———. "The Maximum Duration of a Total Solar Eclipse." *Publications of the American Astronomical Society* 6 (1931): 265–266.

Little, Robert T. *Astrophotography: A Step-by-Step Approach*. New York: Macmillan, 1986.

Lovell, Bernard, ed. *Astronomy*. The Royal Institution Library of Science. 2 vols. Barking, Essex: Elsevier Publishing, 1970.

Marschall, Laurence A. "A Tale of Two Eclipses." *Sky & Telescope* 57 (Feb. 1979): 116–118.

Maunder, Michael. "Eclipse Chasing." In *Yearbook of Astronomy 1990*, edited by Patrick Moore. New York: W W Norton, 1989.

Meadows, A. J. *Early Solar Physics*. Oxford: Pergamon Press, 1970.

Meeus, Jean, Carl C. Grosjean, and Willy Vanderleen. *Canon of Solar Eclipses*. Oxford: Pergamon Press, 1966. (Solar eclipses from A.D. 1898 to 2510.)

Menzel, Donald H., and Jay M. Pasachoff. *A Field Guide to the Stars and Planets*. 2d ed., rev. and enlarged. Boston: Houghton Mifflin, 1983.

Mitchell, Samuel A. *Eclipses of the Sun*. 5th ed. New York: Columbia University Press, 1951.

Mucke, Hermann, and Jean Meeus. *Canon of Solar Eclipses: –2003 to +2526*. Vienna: Astronomisches Büro, 1983.

Needham, Joseph, and Wang Ling. *Science and Civilisation in China*. Vol. 3, *Mathematics and the Sciences of the Heavens and the Earth*. Cambridge: Cambridge University Press, 1959.

Neugebauer, Otto. *The Exact Sciences in Antiquity*. 2d ed. Providence: Brown University Press, 1957.

Newton, Robert R. *Ancient Astronomical Observations and the Accelerations of the Earth and Moon*. Baltimore: Johns Hopkins University Press, 1970.

Oppolzer, Theodor von. *Canon of Eclipses*. Translated by Owen Gingerich. New York: Dover, 1962. (Solar and lunar eclipses from 1207 B.C. to A.D. 2161.)

Osterbrock, Donald E., John R. Gustafson, and W. J. Shiloh Unruh. *Eye on the Sky: Lick Observatory's First Century*. Berkeley: University of California Press, 1988.

Pepin, R. O., J. A. Eddy, and R. B. Merrill, eds. *The Ancient Sun: Fossil Record in the Earth, Moon and Meteorites*. Proceedings of the Conference on the

Ancient Sun. Boulder, Colorado: October 16–19, 1979. New York: Pergamon Press, 1980.

Rao, Joe. *Your Guide to the Great Solar Eclipse of 1991*. Cambridge, Mass.: Sky Publishing, 1989.

Sébillot, Paul Y. *Le folk-lore de France*. T. 1, *Le ciel et la terre*. Paris: Librairie orientale & américaine, 1904.

Silverman, Sam, and Gary Mullen. "Eclipses: A Literature of Misadventures." *Natural History* 81 (June/July 1972): 48–51, 82.

Stegemann, Viktor. "Finsternisse." In *Handwörterbuch des Deutschen Aberglaubens*, edited by Hanns Bächtold-Stäubli. Bd. 2. Berlin: W. de Gruyter, 1930. Columns 1509–1526.

Stephenson, F. Richard, and David H. Clark. *Applications of Early Astronomical Records*. Monographs on Astronomical Subjects, no. 4. New York: Oxford University Press, 1978.

Sweetsir, Richard A., and Michael D. Reynolds. *Observe: Eclipses*. Washington, D.C.: Astronomical League, 1979.

Thompson, J. Eric S. *A Commentary on the Dresden Codex: A Maya Hieroglyphic Book*. Philadelphia: American Philosophical Society, 1972.

Todd, Mabel Loomis. *Total Eclipses of the Sun*. Rev. ed. Boston: Little, Brown, 1900.

Zirker, Jack B. *Total Eclipses of the Sun*. New York: Van Nostrand Reinhold, 1984.

Index

Bold numerals in the index indicate a glossary entry, and italic numerals indicate a diagram or an illustration.